Tapering and Peaking for Optimal Performance

Iñigo Mujika, PhD

University of The Basque Country

Human Kinetics

Library of Congress Cataloging-in-Publication Data

Mujika, Iñigo.
  Tapering and peaking for optimal performance / Iñigo Mujika.
     p. cm.
  Includes bibliographical references and index.
  ISBN-13: 978-0-7360-7484-1 (soft cover)
  ISBN-10: 0-7360-7484-8 (soft cover)
  1. Athletes--Training of.   2. Physcial education and training.   3.
Sports--Physiological aspects.   4. Sports--Psychological aspects.   I.
Title.
  GV711.5.M85 2009
  613.7'11--dc22

                                            2008054200

ISBN-10: 0-7360-7484-8 (print)   ISBN-10: 0-7360-8545-9 (Adobe PDF)
ISBN-13: 978-0-7360-7484-1 (print)   ISBN-13: 978-0-7360-8545-8 (Adobe PDF)

Selected text previously published by the author in *Journal of Sports Sciences* 27(3), pages 195-202; *Medicine & Science in Sports & Exercise* 28(2), pages 251-258, and 35(7), pages 1182-1187; *Olympic Coach* 18(4), pages 9-11; and *Sports Medicine* 34(13), pages 891-927, is included with permission.

The Web addresses cited in this text were current as of November 2008, unless otherwise noted.

Acquisitions Editor: John Dickinson; **Developmental Editor:** Elaine H. Mustain; **Managing Editor:** Melissa J. Zavala; **Assistant Editor:** Christine Bryant Cohen; **Copyeditor:** Julie Anderson; **Proofreader:** Pamela Johnson; **Indexer:** Nancy Ball; **Permission Manager:** Dalene Reeder; **Graphic Designer:** Nancy Rasmus; **Graphic Artist:** Yvonne Griffith; **Cover Designer:** Keith Blomberg; **Photographer (cover):** © Quinn Rooney/Getty Images Sport; **Photographer (interior):** © Human Kinetics, unless otherwise noted; **Photo Asset Manager:** Laura Fitch; **Photo Production Manager**: Jason Allen; **Art Manager:** Kelly Hendren; **Associate Art Manager:** Alan L. Wilborn; **Illustrator:** Keri Evans; **Printer:** United Graphics

Printed in the United States of America   10 9 8 7 6 5 4 3 2 1

The paper in this book is certified under a sustainable forestry program.

Human Kinetics
Web site: www.HumanKinetics.com

United States: Human Kinetics
P.O. Box 5076
Champaign, IL 61825-5076
800-747-4457
e-mail: humank@hkusa.com

Canada: Human Kinetics
475 Devonshire Road Unit 100
Windsor, ON N8Y 2L5
800-465-7301 (in Canada only)
e-mail: info@hkcanada.com

Europe: Human Kinetics
107 Bradford Road
Stanningley
Leeds LS28 6AT, United Kingdom
+44 (0) 113 255 5665
e-mail: hk@hkeurope.com

Australia: Human Kinetics
57A Price Avenue
Lower Mitcham, South Australia 5062
08 8372 0999
e-mail: info@hkaustralia.com

New Zealand: Human Kinetics
Division of Sports Distributors NZ Ltd.
P.O. Box 300 226 Albany
North Shore City
Auckland
0064 9 448 1207
e-mail: info@humankinetics.co.nz

Nire gurasoei—To my parents

contents

part I   Scientific Bases of Tapering   1

part II   Tapering and Athletic Performance   69

High level performance may seem easy to fans of sport; athletes and coaches know that, in fact, it is very difficult to achieve. Furthermore, maintaining high level performance over a full season, several years, or a complete sport career is next to impossible.

Winning five Tour de France titles in consecutive years is an achievement in which I take great pride. Obviously, I was fortunate to have a body to perform at such a high level, and the mental toughness to train and practice to maximize that physical potential. But I also benefited from the research and knowledge of others as to how best to prepare for major competitions.

Now even better and more extensive training information is available to athletes. *Tapering and Peaking for Optimal Performance* is one of the finest examples of this. Author Iñigo Mujika, one of the world's top experts on how to taper workouts prior to events to get the very best results in competition, presents the most recent research-based information on the topic in a way that can be understood and used by students, coaches, and athletes, not just scientists.

Prior to each Tour, I scaled back my training just the right amount to ensure that my body and my mind would be fresh and ready to go at the start of the first stage. Getting an edge in the time trials required great speed and power. Keeping the yellow jersey meant having sufficient endurance to maintain or extend that lead in the mountains. Effective tapering allowed me to do both.

You'll find similar stories from world champion athletes and coaches in *Tapering and Peaking for Optimal Performance*. As a group we provide pretty convincing anecdotal evidence that what Dr. Mujika prescribes works, not just in the lab but in the most demanding sport competitions. I recommend that you read and use this book to achieve the results you're seeking.

Miguel Indurain
Winner Tour de France 1991, 1992, 1993, 1994, 1995
Winner Giro D'Italia 1992, 1993
Individual Time Trial World Champion 1995
Individual Time Trial Olympic Champion 1996

The taper that immediately precedes a major competition is without doubt one of the most important phases of the training season. The temporal proximity between the taper and competition makes the former a key performance-determining factor within a training plan. A well-designed taper will allow an athlete to display, when it really counts, all the physiological, psychological, and performance adaptations that he or she has accumulated through many months of hard training; a successful taper can even further enhance those adaptations. Conversely, an inadequate tapering plan can make all previous training useless, seriously jeopardizing an athlete's possibilities of competitive success.

How This Book Can Help You

Most athletes, coaches, and sport scientists are aware of the key role of the taper in the preparation for competition, but few are certain about the most suitable tapering strategies for their individual needs. In fact, this is usually the training element about which coaches feel most insecure: When should the taper start? How much should the training load be reduced? Which is the most efficient tapering method? Will the athletes detrain instead of peaking their performance? Because there are no simple answers to these questions, the taper has often been planned and designed following trial and error. As a consequence, a range of tapering strategies are used by athletes and their support teams with a view to optimizing sport performance, although it is clear that the most efficient methods to prepare athletes for elite-level competition are those based on proven scientific principles.

This book compiles for the first time the available body of scientific data on tapering; its physiological and psychological effects; how these effects relate to athletic performance; and the experience-based practical knowledge of some of the world's most successful athletes and coaches.

A unique feature of this book is that we have designed it to address the concerns of scientists and scholars, elite athletes and coaches, and all those athletes and coaches in between who want to improve their performance. We have done this by providing a scientific text in parts I and II as the basis for the information. But to that highly technical text, we have added several special elements to enable those who are not interested in the scientific details to understand the basis and conclusions of the technical information. These include

- informational special elements that explain concepts;
- a glossary that defines terms in nonscientific language;
- At a Glance special elements that periodically summarize the scientific text, again in language that should be clear to any athlete or coach of any level who is interested in gaining a competitive advantage through increasing athletic performance; and
- a thorough index that includes page numbers not only for general concepts but for every term that is defined in the glossary.

Parts I and II of the book describe relevant scientific investigations in a great deal of detail, both in the text and in figures and tables, and all the studies that are mentioned are fully referenced, so the academic reader will not be disappointed. However, the book also contains many features to make it fully accessible to nonscientists, such as a

glossary and explanatory At a Glance elements. These elements use nontechnical vocabulary to summarize and interpret the scientific material and include advice on how to apply the scientific conclusions to the art of coaching- and training-program design to achieve optimal performance with the taper. Moreover, part III is written entirely by elite athletes and coaches and narrates their personal experiences with tapering, including in many instances their training programs and tips that they have found to be particularly important for their own performance or coaching. These narratives do not use technical jargon and are accessible to any who are interested in how to use the taper in their own training and competition.

What You Will Learn From This Book

Tapering and Peaking for Optimal Performance is divided into three major parts, in which scientific, performance, and practical issues are analyzed in depth.

Part I, Scientific Bases of Tapering, contains four chapters that explain what tapering is and the physiological and psychological changes that it brings about. Chapter 1 defines the concept of tapering and describes its aims using data from the scientific literature. These data suggest that a successful taper should allow athletes not only to recover their full physiological and performance potential but also to enhance this potential. Chapter 1 also describes the different tapering models and designs that are used by coaches and athletes throughout the world. Chapter 2 deals with some of the physiological changes associated with the taper. It presents a comprehensive compilation of cardiac, respiratory, and metabolic effects of the taper. Chapter 3 completes the overview of physiological changes that take place during the taper, by analyzing biochemical, hormonal, muscular, and immunological changes that have been reported in relation to tapering programs. Chapter 4 describes how a taper can affect the psychological state of athletes in the lead-up to competition. The chapter also presents psychological training tips for athletes involved in a tapering phase. The description of physiological and psychological changes associated with the taper establishes a framework for sport scientists, students, coaches, and athletes to better understand the performance implications of a tapering program.

Part II, Tapering and Athletic Performance, includes four chapters devoted to the performance implications of the taper. Chapter 5 describes how reducing the different training variables (i.e., intensity, volume, and frequency) and varying the duration of the taper may affect athletic performance. The impact of environmental factors such as jet lag, heat, and altitude on tapering design is also discussed. In chapter 6, observed performance gains associated with the taper in various sports are described, and these gains are put into perspective by relating them to competition outcomes. Chapter 7 provides a detailed description of the knowledge gained from mathematical models relating to taper design. The chapter also discusses the possibilities for further insight into tapering strategies through computer simulations based on mathematical modeling, as well as limitations to this approach. Chapter 8 deals with the unique aspects of tapering and peaking for team sports, summarizing the scientific knowledge on tapering for a regular season or for a major tournament such as the Olympic Games. The information included in part II is key for designing the most effective taper plan for a particular athlete or team in a specific situation, setting realistic performance goals for competition, and avoiding negative outcomes that could be associated with a deficient taper program.

Part III, Elite Sports Figures on Tapering and Peaking, contains four chapters dealing with practical aspects of tapering. All four chapters gather ideas, anecdotal evidence, and detailed training information provided by successful coaches and athletes from around the world on their tapering programs in the lead-up to major sporting achievements. Chapter 9 compiles practical examples from coaches and athletes involved in endurance sports with different modes of locomotion (swimming, running, triathlon).

Chapter 10 presents similar information relating to speed and power events, including sprint swimming, sprint cycling, sprint running, and gymnastics. Chapter 11 provides information pertaining to precision sports like archery and golf. Chapter 12 deals with team sports, presenting the ideas and programs of successful coaches in field hockey, men's and women's water polo, and rugby. The information included in part III adds practical value to parts I and II. I expect that part III will be particularly appreciated by coaches, athletes, and students.

The purpose of this book is to summarize the scientific evidence and experience-based anecdotal knowledge on the physiological, psychological, and performance consequences of various tapering strategies, with the aim of helping athletes, coaches, and sport scientists face this important time of the season with increased security and confidence. *Tapering and Peaking for Optimal Performance* does not intend to answer all questions regarding the taper, but I hope that the information presented will be of interest to the readers and will contribute to their knowledge and their ability to apply this knowledge in a practical athletic setting.

acknowledgments

In writing this book, I have been fortunate to interact with a lot of colleagues and friends, sport physiologists, coaches, and athletes, whose help I gratefully acknowledge. I am particularly indebted to the athletes and coaches who contributed their time, knowledge, and experience to complete part III of the book. I am sure it is these contributions that make this book special. The following people were instrumental in making contact with some of the contributors to part III: Unai Castells, Xabier De la Fuente, Jackie Fairweather, Warwick Forbes, Andrew "Scott" Gardner, Sergio Gómez, Peter Hespel, Bojan Matutinovic, Timothy Noakes, and Cathy Sellers.

Thanks are also due to Pruden Indurain for his help in getting the forward organized; and of course, to the great cycling champion, Miguel Indurain, for doing me the great honor of writing it. Thanks to Jesús Flores, Daniel Márquez, and Darío Rodríguez of Finisher Triatlón magazine for their permission to use the part opener photographs; and to triathletes Hektor Llanos, Eneko Llanos, and Ainhoa Murua for allowing me to use their image in those photographs, and for the great privilege and amazing experience that working with them all these years has been for me.

I would also like to thank Human Kinetics for their interest in *Tapering and Peaking for Optimal Performance* when it was nothing but a project and for all their support to make the project come true. Editorial advice from John Dickinson, Elaine Mustain, and Melissa Zavala has been particularly helpful throughout the preparation of the book.

Finally, I would like to warmly thank Sara for her help, care, understanding, patience, and support throughout the long process of writing this book.

Parts of this book first appeared in the form of scientific articles, which have been compiled, updated, and adapted.

part I

Scientific Bases of Tapering

HEKTOR LLANOS emerges from the ocean at the head of the pack at the 2007 Lanzarote Ironman in the Canary Islands, Spain. Llanos placed 7th in a field of more than 1,000, despite a derailleur malfunction that resulted in several stops and ultimately forced him to ride the majority of the 180 km bike leg of the triathlon in only one gear. Llanos' coach is Iñigo Mujika.

Daniel Márquez

Because the taper is the final phase of training prior to a major competition, it is of paramount importance to an athlete's performance and the outcome of the event. However, there is no training phase during which coaches are more insecure about the most suitable training strategies for each athlete, and they have often relied almost exclusively on a trial-and-error approach. Although this approach has resulted in many innovative training strategies in the past, it also presents two major limitations, namely the ample possibility of error inherent in this hit-or-miss technique, and the imprecision of a "shotgun approach" by which several training-related factors are simultaneously changed in an attempt to obtain the ideal training program. But in case of success, it becomes almost impossible to identify the real reason behind that success (Hawley and Burke 1988). A more scientific approach to planning a tapering program, based on our knowledge about the most important factors behind effective tapering strategies and athlete adaptations, should improve the success of individual programs. The four chapters included in part I of this book present

- a thorough description of the scientific knowledge about tapering strategies and models used by sport scientists, coaches, and athletes around the world;
- the physiological and psychological effects that the taper usually has on an athlete; and
- how these effects relate to tapering-induced changes in competition performance.

Readers of part I will gain a solid understanding of what tapering is and how it affects the athlete's body and mind. With this knowledge, readers will be able to design the most efficient taper programs for optimal performance.

chapter 1

Basics of Tapering

Athletes, coaches, and sport scientists around the world are increasingly pushing the limits of human adaptation and training loads with the aim of achieving top performances at major sport competitions. In many competitive events, these top performances are associated with a marked reduction in the athletes' training load during several days prior to the competition. This segment of reduced training is generally known as the *taper* (Mujika and Padilla 2003a).

Understanding what tapering is and what it implies in terms of training strategy and training contents is the initial step toward improving the quality of the tapering programs designed by coaches and performed by athletes. Defining the taper is therefore a good starting point in this journey to optimal performance.

In the past few decades, the taper has been defined in various ways by researchers and practitioners working with athletes in different parts of the world. Here is a chronological sample:

- A decrease in work level that the competitive swimmer undergoes during practice in order to rest and prepare for a good performance (Yamamoto et al. 1988)

- A specialized exercise training technique that has been designed to reverse training-induced fatigue without a loss of the training adaptations (Neary et al. 1992)

- An incremental reduction in training volume for 7 to 21 days prior to a championship race (Houmard and Johns 1994)

- A progressive, nonlinear reduction of the training load during a variable amount of time that is intended to reduce the physiological and psychological stress of daily training and optimize sport performance (Mujika and Padilla 2000)

- A segment of time when the amounts of training load are reduced before a competition in an attempt to peak performance at a target time (Thomas and Busso 2005)

- A time of reduced training volume and increased intensity that occurs prior to a competition (McNeely and Sandler 2007)

A few key points emerge from a close look at these definitions, including the necessity to decrease the training load, the idea of allowing an athlete to rest so that accumulated fatigue can be reduced, the "shape" and the duration of reduced training, and the final goal of any taper program, which is of course to enhance an athlete's performance in a given competition. These concepts are important because they have major implications for an efficient design of tapering strategies.

Newly available scientific knowledge indicates that a taper is—or should be—a time of reduced training load in the lead-up to a major competition; the load should be reduced progressively, mainly at the expense of training volume and for a variable segment of time that depends on individual profiles of adaptation. The taper is intended to reduce accumulated physiological and psychological fatigue, enhance training adaptations, and optimize sport performance.

AT A GLANCE **Planning a Taper**

Key points to consider when planning a taper are these:

- Reduction of the training load
- Management of fatigue and physiological adaptations
- Type of taper (i.e., taper mode)
- Taper duration
- Performance goals

The most up-to-date scientific and practical knowledge about these key points is discussed in depth in various chapters of this book. Indeed, when research-based, widely accepted scientific training principles are combined with the experience-based practical knowledge of successful coaches and athletes, sport-specific training recommendations for success often emerge (Hawley and Burke 1988).

Aims of the Taper

The previously cited definitions of the taper indicate that the main aim of this training phase is to reduce the negative physiological and psychological impact of daily training. In other words, a taper should eliminate accumulated or residual fatigue, which translates into additional fitness gains. To test this assumption, Mujika and colleagues (1996a) analyzed the responses to three taper segments in a group of national- and international-level swimmers by means of a mathematical model, which computed fatigue and fitness indicators from the combined effects of a negative and a positive function representing, respectively, the negative and positive influence of training on performance (figure 1.1). As can be observed in figure 1.1, NI (negative influence) represents the initial decay in performance taking place after a training bout and PI (positive influence) a subsequent phase of supercompensation.

The mathematical model indicated that performance gains during the tapering segments

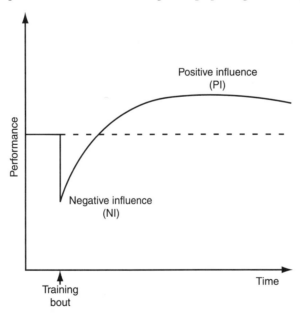

Figure 1.1 Schematic representation of the negative influence (NI) and the positive influence (PI) of training on performance after a training bout.

Adapted, by permission, from I. Mujika, T. Busso, L. Lacoste, et al., 1996, "Modeled responses to training and taper in competitive swimmers," *Medicine & Science in Sports & Exercise* 28: 254.

were mainly related to marked reductions in the negative influence of training, coupled with slight increases in the positive influence of training (figure 1.2). The investigators suggested that athletes should have achieved most or all of the expected physiological adaptations by the time they start tapering, eliciting improved performance levels as soon as accumulated fatigue fades away and performance-enhancing adaptations become apparent.

The conclusions of Mujika and colleagues (1996a), drawn from real training and competition data from elite athletes but attained by mathematical procedures, were supported by several biological and psychological findings extracted from the scientific literature on tapering. For instance, in a subsequent study on competitive swimmers, Mujika and colleagues (1996d) reported a significant correlation between the percentage change in the testosterone–cortisol ratio and the percentage performance improvement during a 4-week taper (figure 1.3). Plasma concentrations of *androgens* and *cortisol* have been used in the past as indexes of anabolic and catabolic tissue activities, respectively (Adlercreutz et al. 1986). Given that the balance between *anabolic* and *catabolic hormones* may have important implications for recovery processes after intense training bouts, the testosterone–cortisol ratio has been proposed and used as a marker of training stress (Adlercreutz et al. 1986, Kuoppasalmi and Adlercreutz 1985). Accordingly, the observed increase in the testosterone–cortisol ratio during the taper would indicate enhanced recovery and elimination of accumulated fatigue. This would be the case regardless of whether the increase in

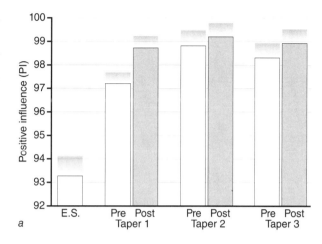

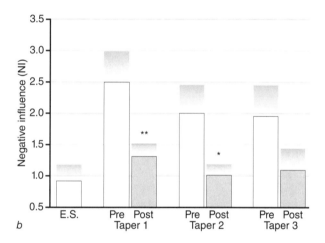

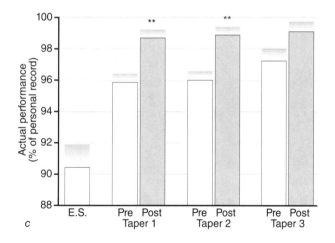

Figure 1.2 Early season (E.S.), pretaper, and posttaper means (± SE) of *(a)* positive influence of training on performance (PI), *(b)* negative influence of training on performance (NI), and *(c)* actual performance. Significant differences were found between pretaper (*$p < .01$) and posttaper (**$p < .01$) values. NI and PI are expressed in the same unit as that used for performance.

Reprinted, by permission, from I. Mujika, T. Busso, L. Lacoste, et al., 1996, "Modeled responses to training and taper in competitive swimmers," *Medicine & Science in Sports & Exercise* 28: 257.

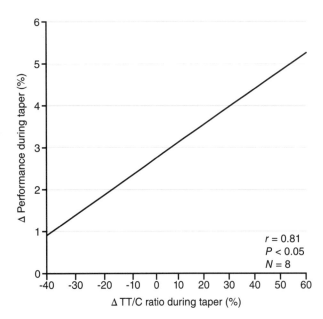

Figure 1.3 Relationship between the percentage improvement in swimming performance and the percentage changes in the total testosterone–cortisol ratio (TT:C) during 4 weeks of tapering in a group of elite swimmers.

With kind permission from Springer Science + Business Media: *European Journal of Applied Physiology*, "Hormonal responses to training and its tapering off in competitive swimmers: Relationships with performance," vol. 74, 1996, p. 364, I. Mujika, J.C. Chatard, S. Padilla, et al., figure 3 (top panel).

the testosterone–cortisol ratio was the result of a decreased cortisol concentration (Bonifazi et al. 2000, Mujika et al. 1996c) or an increased *testosterone* concentration subsequent to an enhanced pituitary response to the preceding time of intensive training (Busso et al. 1992, Mujika et al. 1996d, Mujika et al. 2002a).

Other biological indexes of reduced training stress and increased recovery have been reported in the literature as a result of tapering segments. Different research groups have shown increments in red cell volume, hemoglobin levels, and hematocrit as a result of the taper (Shepley et al. 1992, Yamamoto et al. 1988), and these hematological indexes have been shown to be related with taper-induced performance improvements (Mujika et al. 1997). In line with these results, serum haptoglobin has been shown to increase significantly during the taper (Mujika et al. 2002a). *Haptoglobin* is a *glycoprotein* that binds free hemoglobin released into the circulation to conserve body iron. Because of a rapid removal of the haptoglobin–hemoglobin complex from the blood by the liver, its levels are often below normal in highly trained endurance athletes, suggesting a chronic *hemolytic condition* (Selby and Eichner 1986) that would be reversed during the taper. Increased reticulocyte counts have also been observed at the end of tapering segments in middle-distance runners (Mujika et al. 2000, Mujika et al. 2002a). Taken as a whole, the results indicate that tapering segments in trained subjects are associated with a positive balance between *erythropoiesis* and *hemolysis* in response to the reduced training stress (Mujika and Padilla 2003a).

Androgens and Testosterone

Androgens are natural or synthetic compounds that stimulate or control the growth, development, and maintenance of male sex characteristics. Testosterone is the predominant testicular androgen or male sex hormone, essential for spermatogenesis as well as for growth, maintenance, and development of reproductive organs and secondary sexual characteristics of males.

Among other biological markers, blood levels of *creatine kinase* have also been used as an index of training-induced physiological stress. Creatine kinase concentration in the blood usually decreases in highly trained athletes as a result of the reduced training load that characterizes the tapering segments (Millard et al. 1985, Yamamoto et al. 1988). These and other biological changes associated with the taper are discussed in chapter 3.

Tapering phases are often associated with performance-enhancing psychological changes such as reduced perception of effort, reduced global mood disturbance, reduced perception of fatigue, and increased vigor (Hooper et al. 1999, Morgan et al. 1987, Raglin et al. 1996). The taper has also been associated with an improvement in the quality of sleep in competitive swimmers (Taylor et al. 1997). These psychological changes, further analyzed in chapter 4, can be interpreted as indexes of enhanced recovery from the daily training stress.

It has been proposed that the changes associated with the taper are the result of both a restoration of previously impaired physiological capacities and an enhanced capacity to tolerate training and respond effectively to the training undertaken during the taper (Mujika et al. 2004). In a theoretical study of taper characteristics to optimize performance, Thomas and Busso (2005) examined the training factors that could affect taper efficiency using simulations from a nonlinear mathematical model of the training effects on performance (see also chapter 7). Training responses were simulated from data obtained in a group of subjects involved in a laboratory training program. Thomas and Busso (2005) concluded that simply reducing accumulated or residual fatigue, without simultaneously enhancing adaptations through the training performed during the taper, could be insufficient to reach the best possible performance by the end of the taper. Therefore, adjustment of a taper program should correspond to the optimal compromise between the reduction of fatigue and the continuation, not just the preservation, of positive training adaptations (Thomas and Busso 2005).

Certain hormonal, hematological, biochemical, and psychological markers indicate reduced stress. These are as follows:

- **Hormonal:** Increased testosterone, decreased cortisol, and increased testosterone–cortisol ratio
- **Hematological:** Increased red cell volume, increased hematocrit, increased hemoglobin, increased haptoglobin, and increased reticulocytes
- **Biochemical:** Decreased blood creatine kinase
- **Psychological:** Decreased perception of effort, decreased mood disturbance, decreased perception of fatigue, increased vigor, and improved quality of sleep

---

Aims of the Taper

AT A GLANCE

The performance enhancement that usually takes place with the taper is related to recovery of physiological capacities that were impaired by past training and to restoration of the tolerance to training, resulting in further adaptations during the taper (Mujika et al. 2004, Thomas and Busso 2005). Or, to put it another way, the key objectives for an effective taper are to

- maximally reduce accumulated physiological and psychological stress of daily training and
- restore training tolerance and further enhance training-induced adaptations.

Tapering Models

The term *taper* is now well known by all those who prepare elite athletes for competition, and it is widely used throughout the world in reference to the final training phase leading up to a major race or competition event. But is everyone who speaks of tapering talking about the same training concept? The scientific literature indicates they are not. Houmard (1991) clearly differentiated between the concepts of *reduced training* and tapering. This author indicated that reduced training occurs when training duration, frequency, intensity, or some combination of these elements is reduced by a constant degree. During tapering, on the other hand, these variables are decreased in a systematic, progressive fashion (figure 1.4).

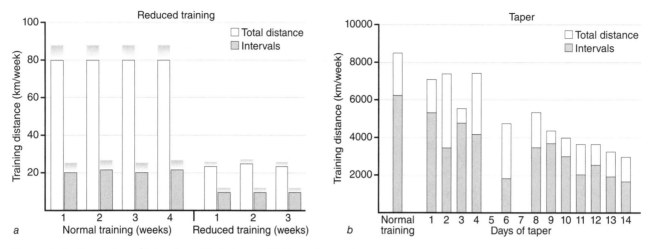

Figure 1.4   Examples of *(a)* reduced training and *(b)* tapering regimens.

Reprinted, by permission, from J.A. Houmard, 1991, "Impact of reduced training on performance in endurance athletes," *Sports Medicine* 12: 381.

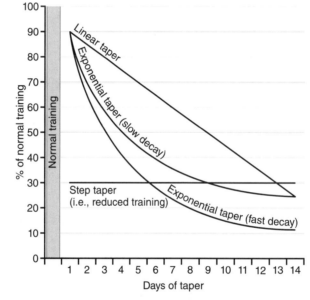

Figure 1.5   The different types of tapers: linear taper, exponential taper with slow or fast time constant of decay of the training load, and step taper (also referred to as reduced training).

Reprinted, by permission, from I. Mujika and S. Padilla, 2003, "Scientific bases for precompetition tapering strategies," *Medicine & Science in Sports & Exercise* 35: 1182-1187.

Four training designs or tapering models have been described and used in the past in an attempt to optimize sport performance. These are shown in figure 1.5.

The training load during the taper is usually reduced in a progressive manner, as implied by the term *taper*. This reduction can be carried out either linearly or exponentially. As shown in figure 1.5, a linear taper usually implies a higher total training load than an exponential taper. In addition, an exponential taper can have either a slow or a fast time constant of decay, the training load being usually higher in the slow decay taper. Nonprogressive standardized reductions of the training load have also been used (figure 1.5). This reduced training procedure, which often maintains but may even improve many of the physiological and performance adaptations gained with training (Mujika and Padilla 2003a), is also referred to as step taper (Banister et al. 1999, Mujika 1998, Zarkadas et al. 1995).

Tapering Models

It is clear that planning a taper implicitly requires reducing the training load. The following three models are most commonly used to achieve this reduction:

- **Linear taper:** the training load is reduced in a systematic, linear fashion
- **Exponential taper:** the training load is reduced in a systematic, exponential fashion, either slowly (slow decay) or suddenly (fast decay)
- **Step taper:** the training load is suddenly reduced by a constant amount

Studies

Sport scientists have made solid efforts to investigate the efficacy of the different tapering models. The most relevant findings eliciting from those efforts are described in this section.

Observational Studies

Some coaches and athletes may have used step tapers to prepare for competition in the past. However, very few studies that have used step tapers in well-trained athletes have been reported in the literature. Martin and colleagues (1994) studied the effects of a 2-week step taper, combining a reduction in training intensity, volume, and frequency, on cycling performance and isokinetic leg strength in a group of well-trained cyclists. These authors found that peak improvements in cycling performance occurred after the first week of taper (8%), whereas improvements in the quadriceps muscle strength (8-9%) peaked at the end of the taper. Houmard et al. (1990a), on the other hand, observed that a 3-week step taper maintained, but did not improve, endurance training adaptations and 5K racing performance in distance runners.

Despite the popularity of both the progressive and nonprogressive approaches to tapering, only one intervention research has compared the performance consequences of these approaches in highly trained athletes. Such an investigation was performed on a group of highly trained triathletes (Banister et al. 1999, Zarkadas et al. 1995). After 1 month of intensive training, the participating triathletes were asked to perform either a 10-day taper in which the training load was reduced exponentially or a step taper of the same duration. The exponential taper brought about a 4.0% improvement in an all-out 5K criterion run and a 5.4% increase in peak power output measured in a ramp cycling test to exhaustion. In contrast, the step taper produced nonsignificant improvements of 1.2% and 1.5%, respectively.

After six additional weeks of intensive training, subjects were asked to perform a 13-day exponential taper, in which the time constant of decay in training volume was either fast (τ = 4 days) or slow (τ = 8 days). The fast exponential taper resulted in 6.3% and 7.9% improvements in the previously mentioned criterion performance measures, whereas improvements with the slow exponential taper were 2.4% and 3.8%. The authors concluded that an exponential taper was a better strategy to enhance performance than a step taper and that the fast-decay protocol (i.e., low-volume taper) was more beneficial to performance than the slow-decay protocol (Banister et al. 1999, Zarkadas et al. 1995).

Results of Observational Studies

Progressive tapering techniques (in which the training load decreases in a stepwise manner) seem to have a more pronounced positive impact on performance than step taper strategies (in which the training load is suddenly reduced by a fixed amount), although step tapers can also improve performance.

Table 1.1 Effect of the Tapering Pattern on Overall Effect Size for Taper-Induced Changes in Performance

Pattern of the taper	Overall effect size, mean (95% confidence interval)	n	p
Step taper	0.42 (–0.11, 0.95)	98	.12
Progressive taper	0.30 (0.16, 0.45)	380	.0001

Data reprinted, by permission, from L. Bosquet, J. Montpetit, D. Arvisais, et al., 2007, "Effects of tapering on performance: A meta-analysis," *Medicine & Science in Sports & Exercise* 39: 1358-1365.

Meta-Analysis

In a recent *meta-analysis* investigation to evaluate the effects of tapering on performance, Bosquet and colleagues (2007) indicated that the way in which the training load is decreased can influence the outcome of a taper, whether it is a single stepwise reduction or a progressive exponential reduction with a fast or eventually a slow decay. According to the reported data, in competitive athletes, maximal gains are obtained with a tapering intervention in which the training volume is exponentially decreased (table 1.1).

Because the pattern of training reduction is not always precisely detailed in the studies included in their meta-analysis, Bosquet and colleagues (2007) gathered linear and exponential tapers together into one single pattern called a *progressive taper*. The results shown in table 1.1 agree thoroughly with the results of Banister and colleagues (1999), as described earlier, who measured higher performance improvements after a progressive taper when compared with a step taper. Similarly, in their study on competitive swimmers, Yamamoto and colleagues (1988) concluded that the observed taper-induced positive physiological changes could be attained in about 7 days of taper, as long as the mean workout distance decreased progressively.

What Is a Meta-Analysis?

A meta-analysis is a statistical procedure that combines the results of previous studies that deal with a set of related research questions. This technique intends to address the limitations of reduced statistical power in studies with small sample sizes. When the results of several studies are pooled together and analyzed descriptively and inferentially, more accurate data analysis is possible and certain hypotheses can be tested.

However, the choice of a given tapering pattern based on the results of Bosquet and colleagues (2007) is complicated by the fact that the majority of available studies used a progressive decrease in training load, resulting in a difference in the number of subjects in the meta-analysis (*n* = 98 for step tapers and *n* = 380 for progressive tapers). Indeed, lower statistical power makes step taper performance gains nonsignificant, even though the overall effect is bigger than that of progressive taper. These authors concluded that future investigations should test the predictions of this meta-analysis with similar subject numbers per tapering pattern.

Another interesting feature of the meta-analysis study by Bosquet and colleagues (2007) is their attempt to identify ideal tapering patterns for different modes of locomotion. Available data allowed these investigators to describe the influence of the tapering pattern in swimming, running, and cycling. As shown in table 1.2, swimmers and runners obtained better performance outcomes when they performed progressive tapers, but mean effect size was bigger with step tapers for cyclists. However, these results should be interpreted with caution, because they arise from small and unequal athlete samples.

Table 1.2 Effect of the Tapering Pattern on Overall Effect Size for Taper-Induced Changes in Swimming, Running, and Cycling Performance

SWIMMING		
Pattern of the taper	**Mean (95% confidence interval)**	***n***
Step taper	0.10 (–0.65, 0.85)	14
Progressive taper	0.27 (0.08, 0.45)*	235
RUNNING		
Pattern of the taper	**Mean (95% confidence interval)**	***n***
Step taper	–0.09 (–0.56, 0.38)	36
Progressive taper	0.46 (0.13, 0.80)*	74
CYCLING		
Pattern of the taper	**Mean (95% confidence interval)**	***n***
Step taper	2.16 (–0.15, 4.47)	25
Progressive taper	0.28 (–0.10, 0.66)+	55

*$p < .01$. +$p < .10$.

Data reprinted, by permission, from L. Bosquet, J. Montpetit, D. Arvisais, et al., 2007, "Effects of tapering on performance: A meta-analysis," *Medicine & Science in Sports & Exercise* 39: 1358-1365.

Bosquet and colleagues (2007) also pointed out that strategies other than simple step, linear, and progressive tapers are being tried by athletes in different sports. One of these alternative strategies consists of an advanced reduction in the training load, followed by a subsequent increase in the lead-up to competition. The rationale behind this tapering design is that the athlete takes advantage of reduced fatigue levels to enhance training tolerance and respond effectively to the training undertaken during the taper, as suggested in the previous section. The relevance of an increase in training load at the end of the taper is corroborated anecdotally by the progressive improvement in performance often observed in an athlete from the first round of a competition to the final. Even though experimental data are not available regarding the suitability of this tapering mode in highly trained athletes, a recent simulation based on mathematical modeling suggests that this may be a suitable alternative to classic tapering designs (see also chapter 7).

Meta-Analysis Results

AT A GLANCE

According to a recent meta-analysis (Bosquet et al. 2007), maximal performance gains can be expected when the training volume is reduced progressively, particularly in swimming and running, but step tapers may be suitable to enhance cycling performance. Alternative strategies such as initially reducing and then increasing the training load at the end of the taper could also be beneficial to performance, but this possibility needs to be experimentally tested.

Other Studies

The purpose of the study by Thomas and colleagues (in press) was to use a nonlinear mathematical model to determine whether a two-phase taper is more effective than a simple progressive taper. Reponses to training were simulated from the model parameters previously determined in competitive swimmers (Thomas et al. 2008). For each participant, the optimal progressive taper after a simulated 28-day overload training

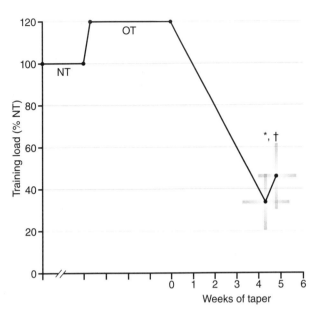

Figure 1.6 Changes in the training load during the normal training segment (NT), the overload training segment (OT), and the optimal two-phase taper in a group of elite swimmers. Values are mean ± SE and are expressed in percentage of NT values. *Significantly greater than the final value of the first phase of the two-phase taper ($p < .05$); †significantly greater than the final value of the optimal linear taper ($p < .05$).

segment was compared with a two-phase taper of the same duration and the same reduction in training load, except for the last 3 days. In the final 3 days, the training load varied in a linear manner to elicit the best possible individual performance. As the authors had hypothesized, the highest performance was achieved after a moderate increase in the training load during the last 3 days of the taper. The optimal variation in the training load during the final 3 days of the two-phase taper was an increase from 35% ± 32% to 49% ± 46% of normal training in the swimmers (figure 1.6).

Interestingly, for one of the swimmers, the optimal training load continued to decrease during the last 3 days of the two-phase taper but decreased less quickly than during the first phase and during the linear taper. This observation is a clear indication that not all athletes respond in the same manner to the training undertaken during the taper, which emphasizes the necessity to individualize the tapering strategies in accord with each athlete's adaptation profile.

During the optimal linear taper, the swimmers' simulated performance increased from the end of the 4-week overload segment by 4.0% ± 2.6%. The maximal performance reached with the optimal two-phase taper was slightly better by 0.01% ± 0.01% than with the optimal linear taper (figure 1.7). This slight difference was explained by a bigger gain in the positive influence of training during the two-phase taper than during the linear taper, given that the negative influence of training (i.e., accumulated or residual fatigue) was removed completely in each participant at the end of both tapers.

The major finding of the model study by Thomas and colleagues (in press) is that a moderate increase in training during the last 3 days of the taper does not appear to be detrimental to competition performance. A benefit of a two-phase taper could be expected in comparison with a simple linear taper because of additional adaptations that did not compromise the removal of fatigue. These findings should also be considered when preparing for competitions consisting of multiple rounds over several days, given that the successive qualifying rounds should increase the training load in the lead-up to the final (Thomas et al. in press).

Another example of an alternative tapering design was reported by Mujika and colleagues (2002b). In an observational investigation on the final preparation of swimmers before the Sydney 2000 Olympic Games, these authors provided a graphical representation of a typical 16-week training cycle prior to the Olympic Games for a member of the Australian Olympic swimming team, consisting of approximately 800 km of swimming (mean 55 km/week; range 40-80 km/week). During the 3-week taper that preceded the Olympic Games, the training volume was reduced systematically from about 10 km to 2 km of swimming per day (figure 1.8a). However, a close look at the daily training volume during the taper (figure 1.8b) shows a saw tapering pattern, alternating between high-volume and low-volume training days.

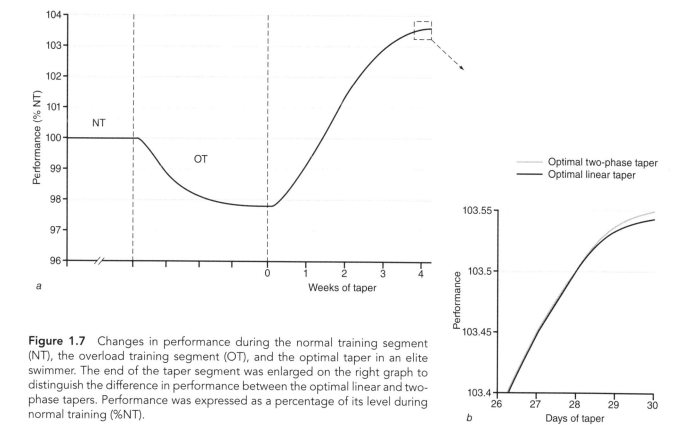

Figure 1.7 Changes in performance during the normal training segment (NT), the overload training segment (OT), and the optimal taper in an elite swimmer. The end of the taper segment was enlarged on the right graph to distinguish the difference in performance between the optimal linear and two-phase tapers. Performance was expressed as a percentage of its level during normal training (%NT).

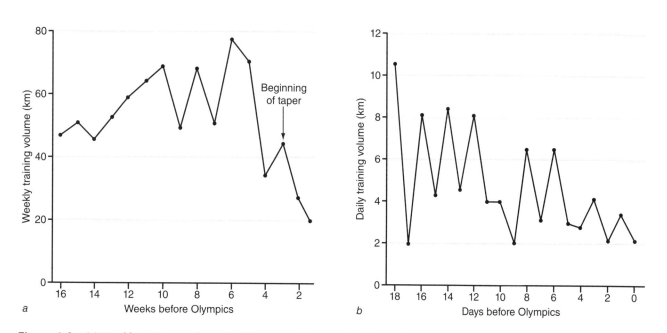

Figure 1.8 (a) Weekly training volume (in kilometers) for a typical Australian swimmer in the 16-week preparation for the Sydney 2000 Olympic Games. (b) Daily training volume (in kilometers) for a typical Australian swimmer in the 18 days prior to the Sydney 2000 Olympic Games.

Reprinted, by permission, from I. Mujika, S. Padilla, and D. Pyne, 2002, "Swimming performance changes during the final 3 weeks of training leading to the Sydney 2000 Olympic Games," *International Journal of Sports Medicine* 23: 583.

Results of Other Studies

A moderate increase in the training load in the days before competition may be beneficial for some, but because not all athletes respond equally to the training undertaken during the taper, tapering strategies must be individualized.

AT A GLANCE **Summary of Research Results**

Research indicates that progressive tapering techniques seem to have a higher positive impact on performance than do step tapering designs. Investigators should examine whether these differences are related to differing total training load and whether different tapering models are better suited for particular sports and individual athletes. In addition, the suitability of alternative, less well known tapering designs to optimize competition performance should be experimentally analyzed and, if proven to be effective, tried by coaches and athletes in different athletic settings.

Chapter Summary

Preparation for a major competition usually finishes with a segment of reduced training known as a taper. When planning a taper, coaches and athletes should consider by how much they need to reduce the training load, for how long, and in what manner (progressively or suddenly) to eliminate accumulated fatigue but continue adaptation to training. Performance gains during the taper are mostly the result of decreasing accumulated fatigue, but research also suggests that further performance-enhancing adaptations can be achieved during the taper.

Performance-enhancing adaptations include a positive hormonal environment (e.g., more testosterone and less cortisol), more red blood cells (i.e., improved oxygen-carrying capacity), less muscle damage, and a better psychological preparedness to face training and competition.

Research suggests that the most efficient way to reduce the training load is to do so progressively, in a gradual manner, rather than suddenly and by a fixed amount. However, the latter technique may also have a beneficial effect on performance, particularly in cycling. Recent theories suggest that progressively reducing the training load and then increasing it in the final days before competition, when the athletes are well rested and thus can carry out quality training, may also be an efficient tapering approach.

It is likely that an ideal tapering model does not exist and that coaches and athletes should adopt different taper designs to suit each athlete's individual adaptation profile.

chapter 2

Taper-Associated Cardiorespiratory and Metabolic Changes

As discussed in chapter 1, in the hopes of improving athletic performance in competition, sport scientists have done a fair amount of research to design different tapering models, describe their aims, and analyze their effects and suitability. Performance improvement is the ultimate goal of an efficient taper, but for that gain to occur physiological changes must take place in an athlete's body.

Unfortunately, the physiological mechanisms underlying the observed performance changes are not yet completely understood. With the exception of two reports by Mujika and colleagues (Mujika 1998, Mujika et al. 2004), published reviews dealing with the physiological changes associated with the taper in athletes date back to the late 1980s and early 1990s (Houmard 1991, Houmard and Johns 1994, Neufer 1989). Nevertheless, considerable efforts by various sport science groups in the past few years have shed light on the physiological changes associated with the taper and the mechanisms responsible for the observed improvements in performance. Some of the most relevant changes that take place during the taper occur at the cardiac, respiratory, and metabolic levels, allowing the athlete to provide the working muscles with more oxygen and fuels and to use them more efficiently. The aim of this chapter is to compile and synthesize current knowledge on tapering-induced cardiorespiratory and metabolic changes in athletes and to assess the possible relationships between these changes and the performance benefits of the taper.

Cardiorespiratory Adaptations

Given the role that the cardiovascular and respiratory systems play during exercise training, they should respond to tapered training with considerable structural and functional changes, despite the relatively short duration of the taper typically performed by well-trained athletes. These changes and their potential effects on sports performance are described next.

Maximal Oxygen Uptake

The most widely used index of cardiovascular and respiratory fitness is the maximal oxygen uptake or $\dot{V}O_2max$. Research has shown that $\dot{V}O_2max$ can increase or remain unchanged during tapering before competition in highly trained athletes. This is a relevant finding, given that a decrease in $\dot{V}O_2max$ during the taper would most likely

What Is $\dot{V}O_2max$?

The use of energy fuels by the body depends on oxygen (O_2) availability and produces carbon dioxide (CO_2) and water. The amount of O_2 and CO_2 used or released by body tissues can be determined by measuring respiratory O_2 and CO_2. By measuring the volume and gas concentrations of the inspired air going into the lungs and the expired air coming out of the lungs, we can calculate the volume of O_2 consumed. The energy required by the body increases with increasing exercise intensity, and so does O_2 uptake ($\dot{V}O_2$). At a given exercise intensity, $\dot{V}O_2$ does not increase any further despite increasing exercise intensity, indicating that the subject has reached his maximal ability to keep increasing his $\dot{V}O_2$. This peak $\dot{V}O_2$ value is known as maximal oxygen uptake or $\dot{V}O_2max$. Successful aerobic (i.e., endurance) performance is associated with a high $\dot{V}O_2max$ and with an athlete's ability to sustain a high percentage of her $\dot{V}O_2max$ for the duration of an endurance event.

indicate a poorly planned tapering strategy in endurance athletes. Many of the relevant studies are described next, and they as well as additional studies are summarized in table 2.1.

Recent investigations have reported $\dot{V}O_2max$ enhancements of 6.0% in cyclists reducing their weekly training volume by 50% during 7 days. This change was concomitant with a 5.4% improvement in a simulated 20 km time trial. On the other hand, neither an increase in $\dot{V}O_2max$ nor a simulated performance gain was observed in cyclists reducing training volume by 30% or 80% during a 7-day taper (Neary et al. 2003a). The same group also reported an increase in $\dot{V}O_2max$ (2.5%) and simulated performance (4.3%) in cyclists who maintained training intensity but reduced training volume. In contrast, cyclists maintaining training volume but reducing intensity only showed statistically nonsignificant improvements in $\dot{V}O_2max$ (1.1%) and simulated performance (2.2%) (Neary et al. 2003b). These results are in agreement with previous reports indicating that training intensity is a key factor for the maintenance or enhancement of training-induced adaptations and optimization of sports performance (Mujika and Padilla 2003a) (see chapter 5).

In line with these results, Jeukendrup and colleagues (1992) reported a 4.5% increase in cyclists' $\dot{V}O_2max$ at the end of 2 weeks of reduced training (i.e., step taper, consisting of a nonprogressive standardized reduction in the training load), accompanied by a 10% higher peak power output and 7.2% faster 8.5 km outdoor time trial. Well-trained triathletes can increase $\dot{V}O_2max$ by 9.1% and criterion laboratory running (1.2-6.3%) and cycling (1.5-7.9%) performances after 2 weeks of taper (Banister et al. 1999, Zarkadas et al. 1995). Margaritis and colleagues (2003) recently observed 3% gains in both $\dot{V}O_2max$ and simulated duathlon performance during a 14-day taper in long-distance triathletes.

However, several investigators have observed unchanged $\dot{V}O_2max$ values as a result of a taper, but this did not preclude athletes from improving their performance (table 2.1). In a study with high school swimmers tapering for either 2 or 4 weeks, both groups improved their swimming time trial performance by 4% to 8% but $\dot{V}O_2max$ was unchanged (D'Acquisto et al. 1992). Shepley and colleagues (1992) reported similar findings of unchanged $\dot{V}O_2max$ values but enhanced treadmill performance in male cross-country and middle-distance runners, as did Houmard and colleagues (1994), who observed a stable $\dot{V}O_2max$ but a 2.8% gain in a 5K treadmill time trial run and a 4.8% longer time to exhaustion in a progressive run on the treadmill in tapered distance runners. Harber and colleagues (2004) saw no changes in $\dot{V}O_2max$ in runners tapering for 4 weeks, but they tended to improve 8 km performance by 1.1% nonetheless. In a study of male cyclists who tapered by reducing their training frequency by 50% for 10 days,

Table 2.1 Effects of the Taper on Maximal Oxygen Uptake ($\dot{V}O_2$max)

Study (year)	Athletes	Taper duration, days	$\dot{V}O_2$max	Performance measure	Performance outcome, %
Van Handel et al. (1988)	Swimmers	20	↔	NR	NR
Houmard et al. (1990a)	Runners	21	↔	5K indoor race	↔
D'Acquisto et al. (1992)	Swimmers	14-28	↔	100 m, 400 m time trial	4.0-8.0 impr
Jeukendrup et al. (1992)	Cyclists	14	↑	8.5 km outdoor time trial	7.2 impr
Shepley et al. (1992)	Runners	7	↔	Treadmill time to exhaustion	6-22 impr
McConell et al. (1993)	Runners	28	↔	5K indoor race	1.2 decl
Houmard et al. (1994)	Runners	7	↔	5K treadmill time trial	2.8 impr
Zarkadas et al. (1995) and ^Banister et al. (1999)	Triathletes	14	↑	5K field time trial run / Incremental maximal test	1.2-6.3 impr / 1.5-7.9 impr
Rietjens et al. (2001)	Cyclists	21	↔	Incremental maximal test	↔
Dressendorfer et al. (2002a,b)	Cyclists	10	↑ slightly	20 km simulated time trial	1.2 impr
Margaritis et al. (2003)	Triathletes	14	↑	30 km outdoor duathlon	1.6-3.6 impr
Neary et al. (2003a)	Cyclists	7	↑	20 km simulated time trial	5.4 impr
Neary et al. (2003b)	Cyclists	7	↑	40 km simulated time trial	2.2-4.3 impr
Harber et al. (2004)	Runners	28	↔	8 km outdoor race	1.1 impr
Coutts et al. (2007b)	Triathletes	14	↔	3 km run time trial	3.9 impr

decl = decline; impr = improvement; NR = not reported; ↑ indicates increased; ↔ indicates unchanged.

Adapted, by permission, from I. Mujika, S. Padilla, D. Pyne, et al., 2004, "Physiological changes associated with the pre-event taper in athletes," *Sports Medicine* 34: 894.

Dressendorfer and colleagues (2002a, 2002b) observed small improvements in $\dot{V}O_2$max (2.5%) and a simulated 20 km time trial (1.2%). Van Handel and colleagues (1988) also reported stable $\dot{V}O_2$max values (65.4 pretaper vs. 66.6 ml·kg^{-1}·min^{-1}, posttaper) in college-aged swimmers (including Olympic medal winners) tapering for 20 days leading up to the U.S. National Championships. Unfortunately, performance outcomes were not reported in this investigation. Similarly, there were no improvements in $\dot{V}O_2$max or peak power output in a study in which trained cyclists underwent a step taper consisting of either continuous training or intermittent training (Rietjens et al. 2001) or in runners during a 3-week (Houmard et al. 1990a, 1990b) or 4-week step taper (McConell et al. 1993).

Collectively, these studies show improved or stable $\dot{V}O_2$max and performance gains after a taper, particularly where training intensity has been maintained (see chapter 5).

Economy of Movement

The *economy of movement* is defined as the oxygen cost of exercise at a given submaximal exercise intensity. Economy has been assessed before and after taper in runners, swimmers, and cyclists. Results of economy studies have been disparate, with the discrepancies probably related to factors such as differences in the training and tapering programs and the caliber of the athletes.

Houmard and colleagues (1994) assessed the *submaximal energy expenditure* in a group of 18 male and 6 female distance runners performing a progressive 7-day run or cycle taper where total training volume was reduced by 85%. The authors reported a 7% (0.9 kcal/min) decrease in calculated submaximal energy expenditure when participants ran at 80% $\dot{V}O_2$peak on a treadmill. This magnitude of improvement in running economy was evident in seven of the eight subjects performing the run taper but not in those performing the cycle taper (figure 2.1). The investigators suggested that an elevation in the muscle's *mitochondrial capacity*, along with neural, structural, and biomechanical factors, could explain improvements in economy with the taper (Houmard et al. 1994). These observations confirmed an earlier investigation by the same group that also described a lower oxygen cost of running after a 3-week segment of reduced training (Houmard et al. 1990a).

More recently, Harber and colleagues (2004) failed to observe any change in running economy at 16 km/hr in university-level cross-country runners following a 4-week taper. This difference may be related to the 25% training volume reduction during the taper, because studies suggest that a 41-60% reduction in training volume is required for a taper to achieve optimal results for most athletes (see chapter 5).

Improvements in economy of movement have also been reported in swimming, but these gains appear to be inversely related to the caliber of the athletes. High school–level male and female swimmers tapering for 2 or 4 weeks showed downward shifts in their $\dot{V}O_2$–velocity curves (i.e., economy) between 4.9% and 15.6% and between 8.5% and 16.7%, respectively, at a range of different swimming velocities (D'Acquisto et al. 1992). The authors suggested that changes in economy were dependent on reductions in training volume and, like Houmard and colleagues (1994), speculated that the taper had a beneficial effect on biomechanics, allowing the swimmers to develop better stroke mechanics (D'Acquisto et al. 1992). Johns and colleagues (1992) also reported declines of 5% to 8% in the oxygen cost of swimming after 10 or 14 days of taper in intercollegiate swimmers. In contrast, Van Handel and colleagues (1988) failed to observe any tapering-induced changes in the economy curves of swimmers of much higher caliber, who were described as considerably more economical than less skilled swimmers.

In cycling, Dressendorfer and colleagues (2002a) failed to observe a marked improvement in economy at a power output of 200 W in male cyclists tapering for 10 days, when values were compared with the previous two training phases characterized by high volume and high intensity, respectively. Nevertheless, the oxygen cost of cycling at that power output remained lower after the taper (2.82 L/min) than at baseline (2.95 L/min). Houmard and colleagues (1989) reported unchanged submaximal $\dot{V}O_2$max after a 10-day taper in runners. Rietjens and colleagues (2001) did not observe any changes in either the oxygen cost of cycling at 165 or 270 W in cyclists before and after a 3-week step taper, nor did McConell and colleagues (1993) after a 4-week step taper in runners at 65%, 85%, or 95% $\dot{V}O_2$max.

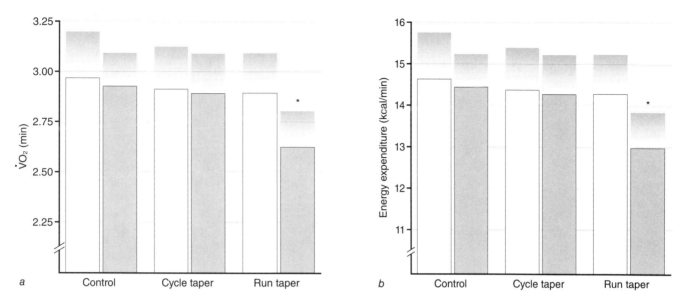

Figure 2.1 (a) Absolute oxygen consumption and (b) calculated energy expenditure during the submaximal treadmill runs (80% $\dot{V}O_2$peak) before (white) and after (gray) the 7-day experimental taper segments in each group. *Significantly different (p < .05) from pre.

J.A. Houmard, B.K. Scott, C.L. Justice, et al., 1994, "The Effects of taper on performance in distance runners," *Medicine & Science in Sports & Exercise* 26(5): 624-631.

Muscle Oxygenation

Neary and colleagues (2005) used near-infrared spectroscopy (NIRS) to examine the effects of a taper on *muscle oxygenation* during cycling. NIRS is a non-invasive optical technique based on the differential absorption properties of hemoglobin and myoglobin in the near-infrared (700-1,000 nm) range. These authors examined muscle oxygenation of the right vastus medialis during a simulated 20 km cycling time trial before and after a 7-day taper consisting of either a 30%, a 50%, or an 80% reduction in weekly training volume. Mean tissue oxygenation during the cycling test was significantly greater (i.e., increased muscle deoxygenation) following the 50% training volume reduction taper (figure 2.2). In addition, there was a moderately high correlation between the change in tissue oxygenation and 20 km

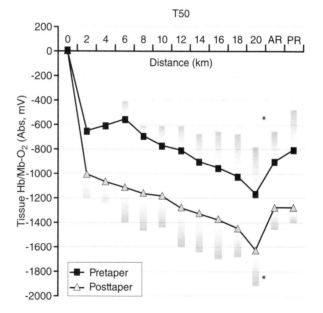

Figure 2.2 Group mean (± SD) tissue oxygenation (Hb/Mb-O_2; absorbency, mV) during the simulated 20 km time trial performance ride pre- and posttaper in the group that reduced weekly training volume by 50%. *Significance (p < .05) between pre- and posttaper trials.

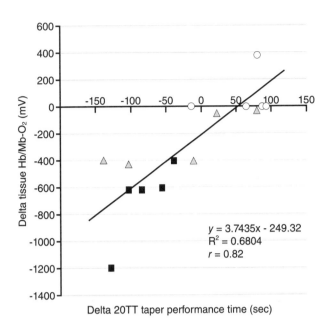

Figure 2.3   Relationship (r = .82) between the change (delta) in tissue Hb/Mb-O$_2$ (mV) versus 20TT taper performance time (minutes) for all subjects combined (N = 11) after tapering. Note that a negative value for performance time indicates an improvement in performance (i.e., faster 20TT ride). Delta scores were determined by posttaper minus pretaper values. Each symbol represents a different taper group (circle T30; square T50; triangle T80).

time trial performance (figure 2.3), which suggests that metabolic changes that occur in the vastus medialis muscle are in part responsible for performance changes at the whole-body level. In other words, the taper induced increased oxygen extraction, which was associated with increased performance (Neary et al. 2005).

Cardiac Function and Morphology

Athletic training implies the activation of most or all of the human body's muscles. When we talk about muscular training, we tend to think of the skeletal muscles but often forget about the most important muscle of all, the cardiac muscle. The heart is deeply involved in all training activities that an athlete performs, no matter what the athletic event, the training intensity, or the duration might be. The heart responds to the training solicitations and adapts by selectively modifying its functions and dimensions. This is true during times of intensive training and also during the segments of taper and peaking in preparation for competition.

Resting Heart Rate

Few reports are available on the effects of tapering on athletes' resting heart rate (HR), but the general consensus of investigators is that resting HR does not appear to change during this phase of training (table 2.2). Haykowsky and colleagues (1998) reported unchanged mean resting HR values of 57 and 59 beats/min before and after tapering for 2 weeks, respectively, at an altitude of 1,848 m and reported 59 and 56 beats/min at 1,050 m. Unchanged resting HR values were also observed by Flynn and colleagues (1994) before and after 3 weeks of taper in collegiate cross country runners (51 vs. 52 beats/min) and collegiate swimmers (54 vs. 55 beats/min). In line with these results, stable resting HRs were observed in a group of international-level swimmers tapering for 2 weeks (Hooper et al. 1999), in elite weightlifters tapering for either 1 or 4 weeks (Stone et al. 1996), and runners undergoing a 4-week step taper (McConell et al. 1993). In contrast with these reports, Jeukendrup and colleagues (1992) described a decrease from 54 to 51 beats/min in the sleeping HRs of their group of cyclists after 2 weeks of reduced training. However, these subjects were purposely overreached before the taper, a condition that may have elicited abnormally high resting HRs (Achten and Jeukendrup 2003).

Table 2.2 Effects of the Taper on Heart Rate (HR)

Study (year)	Athletes	Taper duration, days	Resting HR	Maximal HR	Submaximal HR	Performance measure	Performance outcome, %
Costill et al. (1985)	Swimmers	14	NR	NR	↔	46-1,509 m competition	2.2-4.6 impr
Houmard et al. (1989)	Runners	10	NR	↔	↔	Incremental maximal test	↔
Houmard et al. (1990a)	Runners	21	NR	↑ slightly	↔	5K indoor race	↔
D'Acquisto et al. (1992)	Swimmers	14-28	NR	↓	↔	100 m, 400 m time trial	4.0-8.0 impr
Jeukendrup et al. (1992)	Cyclists	14	↓	↑	↑	8.5 km outdoor time trial	7.2 impr
McConell et al. (1993)	Runners	28	↔	↑ slightly	↔	5K indoor race	1.2 decl
Flynn et al. (1994)	Runners Swimmers	21	↔ ↔	NR NR	↔ NR	Treadmill time to exhaustion 23 m, 366 m time trial	↔ ≈3 impr
Houmard et al. (1994)	Runners	7	NR	↔	↑ slightly	5K treadmill time trial	2.8 impr
Stone et al. (1996)	Weightlifters	7-28	↔	NR	↔	Competition	8.0-17.5 kg impr
Haykowsky et al. (1998)	Swimmers	14	↔	NR	NR	NR	NR
Hooper et al. (1999)	Swimmers	14	↔	↑ slightly	NR	100 m time trial	↔
Martin and Andersen (2000)	Cyclists	7	NR	↑ slightly	↔	Incremental maximal test	≈6 impr
Rietjens et al. (2001)	Cyclists	21	NR	↔	↔	Incremental maximal test	↔
Dressendorfer et al. (2002a)	Cyclists	10	NR	NR	↔	20 km simulated time trial	1.2 impr
Neary et al. (2003a,b)	Cyclists	7	NR	NR	↔	20 km simulated time trial	5.4 impr

decl = decline; impr = improvement; NR = not reported; ↓ indicates decreased; ↑ indicates increased; ↔ indicates unchanged.

Reprinted, by permisison, from I. Mujika, S. Padilla, D. Pyne, et al., 2004, "Physiological changes associated with the pre-event taper in athletes," *Sports Medicine* 34: 897.

Overreaching Versus Overtraining

According to Halson and Jeukendrup (2004), athletes experience minor fatigue and acute reductions in performance as a result of the normal training process. When an imbalance between training stress and recovery takes place, it is thought that overreaching and possibly overtraining may develop. Overreaching normally occurs as a result of intensified training, and it could be considered a normal training outcome for elite athletes. Recovery from overreaching should take a relatively short time (several days or weeks), and this recovery should result in a supercompensatory effect. Overtraining, on the other hand, should be considered a pathological outcome of long segments of excessive training and insufficient recovery, resulting in a long-term reduction of performance capacity. The time needed to recover from the overtraining syndrome can be months or years.

Maximal Heart Rate

Results from investigations addressing the effects of taper on maximal HR are not consistent, and values have variously been shown to decrease, remain constant, or increase after a taper (table 2.2). For instance, D'Acquisto and colleagues (1992) reported lower maximal HR in swimmers after tapers lasting two (187 vs. 192 beats/min) or four (185 vs. 194 beats/min) weeks. Maximal HR did not change during taper in runners performing a progressive treadmill run to exhaustion (Houmard et al. 1994, Houmard et al. 1989) or in a group of cyclists performing an incremental cycling test to exhaustion (Rietjens et al. 2001). In contrast, maximal HR was slightly increased after 1 week of taper in cyclists (Martin and Andersen 2000), 2 weeks of taper in swimmers (Hooper et al. 1999), and 3 (Houmard et al. 1990a) and 4 (McConell et al. 1993) weeks of reduced training in runners. In an overreaching–recovery protocol, maximal HR increased from 178 to 183 beats/min, but this posttaper value was similar to that measured before undertaking the intensified training (Jeukendrup et al. 1992). Reduced maximal HR after intensified training is related to *catecholamine* depletion and is indicative of the neuroendocrinologic nature of an athlete's overreached or overtrained status (Lehmann et al. 1991; Lehmann et al. 1992). A possible explanation for the inconsistent findings could relate to opposite effects on maximal HR of blood volume expansion and the level of catecholamine depletion that may have been incurred during the preceding phase of intense training. (For a discussion of catecholamines, see page 44.)

Submaximal Heart Rate

Most of the available literature on the effects of taper on submaximal exercise HR indicates few changes across a range of different athletic activities (table 2.2). No change in HR was observed by D'Acquisto and colleagues (1992) when swimmers were required to swim at submaximal velocities ranging from 1.0 and 1.3 m/s before and after 2 or 4 weeks of taper. Costill and colleagues (1985) reported unchanged postexercise HR after an evenly paced 200-yard (182 m) swim at a speed representing 90% of individual season's-best performance after 1 and 2 weeks of taper. A 10-day taper did not elicit a change in HR when participants ran at 265 m/min or 298 m/min (Houmard et al. 1989). Similarly, a 3-week taper did not affect HR of collegiate runners while running at 75% $\dot{V}O_2$max (161 beats/min before taper vs. 163 beats/min after taper) (Flynn et al. 1994), nor did a 3-week step taper at 65% and 85% $\dot{V}O_2$max (Houmard et al. 1990a) or a 4-week step taper at 65%, 85%, or 95% $\dot{V}O_2$max (McConell et al. 1993). Submaximal HR while cycling at a power output of 200 W was unchanged after a 10-day taper in male cyclists when compared with the preceding high-intensity interval training phase, even though HR was lower than at baseline (137 vs. 152 beats/min, respectively), indicating a positive training adaptation (Dressendorfer et al. 2002a). Similarly, no changes in HR at power outputs of 165 and 270 W were reported before and after 3-week continuous or intermittent training step tapers (Rietjens et al. 2001). In line with these results, Martin and Andersen (2000) observed identical HR–power output slopes before and after a 1-week taper in male collegiate cyclists, but these slopes were markedly shifted to the right when compared with baseline values. Neary and colleagues (2003a) reported that HR during a simulated 20 km time trial was identical before and after taper in a group of male cyclists, but the O_2 pulse increased from 23.1 to 24.8 ml O_2/beat in those subjects reducing weekly training volume by 50% during the 7-day taper, who also improved their simulated performance by 5.4%. Submaximal exercise HR values were also unaffected by 1 or 4 weeks of taper in weightlifters (Stone et al. 1996).

In contrast, HR was elevated during an all-out 5K treadmill run after participants performed both a running taper and a cycling taper. Heart rates were also 4 to 6 beats/min higher during a submaximal run at $\dot{V}O_2$peak, but this change did not attain statistical significance (Houmard et al. 1994). The overreaching–recovery protocol of Jeukendrup and colleagues (1992) induced a clear increase in submaximal HR measured during an 8.5 km outdoor cycling time trial after the taper. These changes were thought to be related with the level of *neuroendrocrine fatigue* in the pretaper condition.

Blood Pressure

I am aware of only three published reports on the effects of taper on resting blood pressure (BP), and all of them failed to show any substantial effect of tapering on BP. Flynn and colleagues (1994) reported pre- and posttaper systolic BP values of 112 and 114 mmHg in eight male runners, respectively, and 118 and 116 mmHg in five male swimmers. Diastolic pressures were 73 and 74 mmHg for the runners and 76 and 78 mmHg for the swimmers. In swimmers, Hooper and colleagues (1999) reported modest declines of 3.4% and 2.2% in systolic and diastolic pressures during taper, with standard deviations of 12.5% and 12.2%, respectively. In terms of a power sport, the results of Stone and colleagues (1996) indicated no tapering-induced changes in the resting BP of elite weightlifters.

Cardiac Dimensions

The only study on the effects of a taper on cardiac dimensions assessed the effects of a 2-week swimming taper at moderate altitudes—1,050 m and 1,848 m (Haykowsky et al. 1998). The taper consisted of a progressive 73% reduction in training volume coupled with a slight increase in the percentage of high-intensity training. The investigators observed no marked change in *diastolic* or *systolic cavity* dimensions, *ventricular septal wall* thickness, estimated absolute or relative left ventricular mass, *stroke volume, cardiac output, cardiac index,* or *fractional shortening.* The authors concluded that 3 weeks of altitude training or control training followed by 2 weeks of taper training was not associated with alterations in structural and functional aspects of cardiovascular dynamics (Haykowsky et al. 1998). Although more studies are needed before definite conclusions can be drawn about the effects of the taper on cardiac dimensions, it is unlikely that these dimensions would be affected, in view of the limited duration and the relatively high training demands of taper programs, in contrast with times of complete training stoppage (Mujika and Padilla 2001).

Ventilatory Function

Few investigators have addressed tapering effects on athletes' *ventilatory function.* Peak ventilatory volume during a simulated 20 km cycling time trial was unchanged as a result of a 7-day taper, but the ventilatory equivalent for O_2 declined from 25.5 to 24.0 l/lO$_2$ (l/lO$_2$ means liters of air ventilating the lungs per liters of oxygen consumed) in those cyclists tapering by reducing training volume by 50% (Neary et al. 2003a). Unchanged maximal ventilation was also observed in runners after a 4-week step taper (McConell et al. 1993). Submaximal ventilation during a 10-min treadmill run declined by 5.4% after a 7-day taper in distance runners (Houmard et al. 1994). Similarly, power output at the *ventilatory threshold* increased by 12% (28 W) in cyclists during a high-intensity, low-volume taper eliciting a 4.3% performance improvement over a simulated 40 km cycling time trial and by 8% (19 W) in a low-intensity, high-volume taper that induced a 2.2% simulated performance gain (Neary et al. 2003b). The same group observed a 12% (27 W) improvement in the ventilatory threshold after tapers lasting either 4 or 8 days (Neary et al. 1992). A smaller improvement of 2.9% was reported by Dressendorfer and colleagues (2002a), coupled with a performance gain over a simulated 20 km time trial of only 1.2%.

Hematology

The amount of red blood cells (erythrocytes) in the blood of an athlete can have a great impact on performance during both training and competition, particularly for an endurance athlete. Red blood cell content in the blood can be described as the balance between red cell production (erythropoiesis) and red cell destruction (hemolysis). The higher the red cell count in the blood, the better the oxygen-carrying capacity to the exercising muscles. Although endurance training is most often associated with improved oxygen-carrying capacity, intensive training can also induce a substantial hemolysis. This, along with iron stores that are often compromised because of excessive training or poor nutritional habits, may jeopardize the hematological status of endurance athletes. In this section, the potential benefits of a taper in terms of enhancing an athlete's hematological status and iron stores are discussed.

Hemolysis Versus Erythropoiesis

The taper is a specialized training strategy that can be accompanied by a positive balance between exercise-induced hemolysis and recovery-facilitated erythropoiesis. Intensive athletic training can result in decreased red blood cells, *hemoglobin* concentration, and *hematocrit* (Coutts et al. 2007a, Mujika et al. 2004). These changes have variously been attributed to a *hemodilution* caused by training-induced expanded plasma volume, an imbalance between *hematopoiesis* and *intravascular hemolysis,* or iron deficiency (Hallberg and Magnusson 1984, Kaiser et al. 1989, Weight 1993).

Terminology in Blood Breakdown and Formation

Intravascular hemolysis is the breakdown of red blood cells with release of hemoglobin within the blood vessels, whereas *extravascular hemolysis* refers to the destruction of red blood cells in the spleen and liver. Hematopoiesis refers to the formation of blood, whereas erythropoiesis specifically refers to the production of erythrocytes (i.e., red blood cells).

Taper-induced increases in blood and red cell volume have been reported in highly trained distance runners. Shepley and colleagues (1992) observed a 15% increase in total blood volume after a 7-day high-intensity, low-volume taper in runners. This training-induced *hypervolemia* is possibly associated with an elevation of plasma *renin* activity and *vasopressin* concentration during exercise and a chronic increase in the water-binding capacity of the blood (Convertino et al. 1981, Convertino et al. 1983, Mujika 1998, Wade and Claybaugh 1980). Shepley and colleagues (1992) also observed a 14% increase in red cell volume and a 2.6% increase in hematocrit, possibly the result of an increased hematopoiesis concomitant to a decreased intravascular hemolysis (figure 2.4).

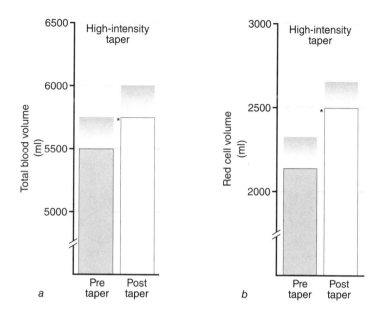

Figure 2.4 Total blood volume *(a)* and red cell volume *(b)* before and after a high-intensity taper ($N = 9$). Values are mean ± SD. *Significant pre- to posttaper differences, $p < .05$.

Reprinted, by permission, from B. Shepley, J.D. MacDougall, N. Cipriano, et al., 1992, "Physiological effects of tapering in highly trained athletes. *Journal of Applied Physiology* 72(Feb): 709. Permission conveyed through Copyright Clearance Center, Inc.

Consistent with the previously described results, hemoglobin concentration and hematocrit increased during the taper in competitive swimmers (Burke et al. 1982a, Rushall and Busch 1980, Yamamoto et al. 1988) and triathletes (Coutts et al. 2007a, Margaritis et al. 2003, Rudzki et al. 1995). These results were attributed to a decreased hemolysis and a net increase in erythrocytes, presumably facilitated by the reduced training load that characterizes tapering (Houmard 1991, Houmard and Johns 1994, Mujika et al. 1997, Neufer 1989, Shepley et al. 1992). This suggestion seems to be confirmed by the findings of Mujika and colleagues (2000), showing a 40% increase in the posttaper *reticulocyte* count in a group of middle-distance runners tapering for 1 week. This change was indicative of an enhanced red cell production resulting in the release of immature erythrocytes, with a smaller hemoglobin content and reduced *mean corpuscular hemoglobin* values (Brodthagen et al. 1985, Mairbäurl et al. 1983, Seiler et al. 1989).

Two additional indexes of a positive red cell balance during taper are increased *serum* haptoglobin and decreased red cell distribution width. Serum haptoglobin is a glycoprotein that binds free hemoglobin released into the circulation to conserve body iron. Lower than normal haptoglobin levels have been found in middle- and long-distance runners and swimmers (Casoni et al. 1985, Dufaux et al. 1981, Pizza et al. 1997, Selby and Eichner 1986). These observations have been attributed to a rapid removal of the haptoglobin–hemoglobin complex from the blood by the liver and suggest a chronic hemolytic condition. In contrast, serum haptoglobin can increase with a 6-day taper in middle-distance runners (Mujika et al. 2002a). This finding was in parallel with a trend toward increased reticulocyte counts, suggesting that the reduced training load undertaken by the athletes during the taper facilitated a positive balance between erythropoiesis and hemolysis.

Plasma, Serum, and Haptoglobin

Are plasma and serum the same thing? Although they are sometimes used as synonyms, plasma and serum are not the same. *Plasma* is the fluid ground substance of whole blood, that is, what remains after the cells have been removed from a sample of whole blood. *Serum* is blood plasma from which fibrinogen and other clotting proteins have been removed as a result of clotting.

What does serum haptoglobin tell us about the balance between red blood cell production and destruction? When high amounts of red blood cells are destroyed, released hemoglobin is captured by the circulating protein haptoglobin, and the two are quickly removed from the blood by the liver; thus, haptoglobin concentration decreases (this occurs during heavy training). When not many red cells are being destroyed, haptoglobin concentration increases because it circulates freely in the blood as there isn't much released hemoglobin to carry to the liver (this occurs during times of light training).

In one study, the red cell distribution width decreased slightly during 12 weeks of intensive training and 4 weeks of taper in highly trained swimmers, to attain significantly lower values at the end of the taper (Mujika et al. 1997). Slight reductions during taper have also been reported in runners and swimmers (Pizza et al. 1997). Decrements in red cell distribution width are considered a positive adaptation to training, given that a high red cell distribution width has been associated with decreased *deformability*, decreased *osmotic resistance,* and increased *mechanical fragmentation* of erythrocytes (Bessman et al. 1983, Kaiser et al. 1989).

With respect to the possible influence of the observed hematological changes on performance, Shepley and colleagues (1992) attributed part of the reported 22% improvement in a treadmill run to fatigue after the taper to the increase in blood and red cell volume. Mujika and colleagues (1997) observed a 2.3% mean competition performance improvement in tapered swimmers, and a positive correlation was found between posttaper red cell count and the percentage improvement in performance attained by the swimmers during taper. Red cell count, hemoglobin, and hematocrit increased by 3.5%, 1.8%, and 3.3%, respectively, in the swimmers for whom the taper was most effective, whereas decreases of 2.2%, 4.3%, and 2.1% occurred in those athletes improving less with the taper (figure 2.5). The authors suggested that the net increase in erythrocyte values observed in the successful swimmers during the taper could have been in part responsible for the higher performance improvement attained, given that small percentage increases in hemoglobin or hematocrit values can result in worthwhile improvements in $\dot{V}O_2$max and exercise capacity (Gledhill 1982, 1985).

Iron Status

Enhanced erythropoietic activity in the bone marrow associated with the taper could jeopardize the iron status of athletes. An iron profile indicative of a prelatent–latent iron deficiency, with normal red cell count and hemoglobin but lowered *ferritin*, serum iron, and *transferrin* saturation and increased transferrin values (Bothwell et al. 1979, Clement and Sawchuk 1984), has been reported in middle-distance runners at the end of 6-day tapers. This, however, did not seem to negatively affect the athletes' competition performance (Mujika et al. 2002a, Mujika et al., 2000). Lowered posttaper ferritin values have also been reported in male cross country runners after a 3-week taper (Pizza et al. 1997) and in triathletes tapering for 2 weeks (Rudzki et al. 1995) but not in swimmers (Mujika et al. 1997).

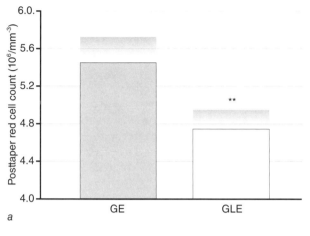

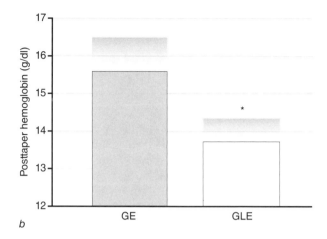

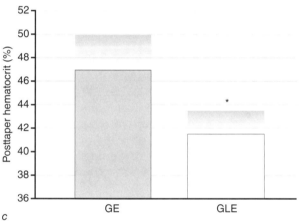

Figure 2.5 Group comparisons for posttaper (a) red cell count, (b) hemoglobin concentration, and (c) hematocrit. GE (group efficient) and GLE (group less efficient) swimmers improving their competition performance with the taper by more than 2% and less than 2%, respectively. Significant differences were found between groups (*$p < .05$ and **$p < .01$).

Reprinted, by permission, from I. Mujika, S. Padilla, A. Geyssant, et al., 1997, "Hematological responses to training and taper in competitive swimmers: Relationships with performance." *Archives of Physiology and Biochemistry* 105(4): 383. Permission conveyed through Copyright Clearance Center, Inc.

Carrying Oxygen to the Muscles During the Taper

AT A GLANCE

Changes in the cardiorespiratory system during the taper may include an increase in maximal oxygen uptake ($\dot{V}O_2$max), and this may contribute to taper-induced performance gains, but enhanced performances can also occur without changes in $\dot{V}O_2$max. A lower oxygen cost of exercise after the taper (i.e., a better economy of movement) can also contribute to improved performances, but this change is more likely to occur in athletes of a lower caliber. Resting, maximal, and submaximal heart rates do not change during the taper, unless athletes show clear signs of overreaching before the taper. Blood pressure, cardiac dimensions, and ventilatory function are not generally affected by the taper, and any observed change seems to be a reflection of the athletes' adaptation to preceding training rather than to the taper itself.

Hematological changes strongly suggest that the reduced training loads undertaken by athletes during the taper facilitate a positive balance between red blood cell production and destruction, contributing to taper-induced performance improvements, but potentially compromising the iron status of the athletes.

Metabolic Adaptations

Energy metabolism underpinning exercise performance can be altered during a preevent taper. Decreases in training load in favor of rest and recovery lower an athlete's daily energy expenditure, potentially affecting energy balance and body composition.

Substrate availability and utilization, blood lactate kinetics, muscle glycogen content, and other metabolic variables may also be altered during the taper.

Energy Balance

Margaritis and colleagues (2003) reported the daily energy intake, energy expenditure, body mass, and body fat of 20 male long-distance triathletes during 4 weeks of overloaded training followed by 2 weeks of tapered training. Energy intake did not change between both training phases (13.8-15.0 vs. 13.2-15.0 MJ/day), whereas energy expenditure decreased from a range of 16.8 to 17.0 MJ/day to a range of 12.1 to 12.7 MJ/day. Total body mass did not change during the taper, but the percentage body fat increased slightly from a range of 11.4% to 11.5% to a range of 11.8% to 12.1% (Margaritis et al. 2003). Similar changes were observed in a 4-week reduced training study on 10 well-trained male distance runners, whose body fat increased from 10.4% to 11.8% (McConell et al. 1993), and in 16 well-trained male triathletes tapering for 2 weeks (sum of nine skinfolds increased from 63.0 to 73.5 mm) (Coutts et al. 2007a). These results indicate that a certain level of muscle mass loss may have taken place during the taper and suggest that athletes tapering for competition should pay careful attention to matching energy intake in accordance with the reduced energy expenditure that characterizes this training interval.

D'Acquisto and colleagues (1992) reported the body mass and percent body fat of female swimmers before and after 2-week and 4-week tapers, and Izquierdo and colleagues (2007) did the same in strength-trained athletes during a 4-week taper. Both groups of researchers observed that neither variable changed significantly. Additional studies have reported stability in body mass in well-trained distance runners during a 3-week step taper (Houmard et al. 1990a, 1990b) and in collegiate cross country runners and swimmers after a 3-week taper consisting of a 20% to 33% weekly reduction in training volume (Flynn et al. 1994). Similarly, nine male cyclists maintained their body mass during 10 days of taper consisting of resting every other day (Dressendorfer et al. 2002a), as did collegiate swimmers preparing for the final meet of the season (Van Handel et al. 1988). These studies, however, did not report the possible changes in fat mass and muscle mass of the athletes.

Substrate Availability and Utilization

There are few reports on the effects of tapering on the *respiratory exchange ratio (RER)*, used as an index of *substrate utilization* during exercise. During submaximal-intensity exercise, RER has been shown to decline or not to change after a taper. A group of club-level cyclists showed a posttaper shift in RER from 0.99 to 0.96 while cycling at a power output of 175 W, suggesting a higher contribution from fat to energy production at moderate exercise intensity (Neary et al. 1992). Houmard and colleagues, on the other hand, reported unchanged RER values during a 10-min continuous run at 80% $\dot{V}O_2$peak (Houmard et al. 1994) and during submaximal running at 265 and 298 m/min (Houmard et al. 1989). Rietjens and colleagues (2001) also showed unchanged rates of fat oxidation during a 90 min steady-state cycle after 7, 14, and 21 days of a step taper. Three- and 4-week step tapers elicited slight but statistically significant increases in runners' RER at 65%, 85%, and 95% $\dot{V}O_2$max (Houmard et al. 1990a, McConell et al. 1993).

During maximal exercise, such as a simulated 20 km cycling time trial, RER values have been shown to remain unchanged after tapering (Neary et al. 2003a). Similarly, unchanged maximal RER values were reported during treadmill running after either a running or a cycling 7-day taper (Houmard et al. 1994). These results suggest that the substrate contribution to power production during maximal intensity exercise is not modified by a taper. This lack of change may be related to stable aerobic–anaerobic work production and oxygen deficit during the taper (Morton and Gastin 1997). Another explanation could relate to increased *muscle glycogen* concentration during the taper, which theoretically induces higher carbohydrate use during both maximal and submaximal exercise. Six days of taper coupled with a high-carbohydrate diet increased carbohydrate oxidation and RER values during cycling at 80% $\dot{V}O_2$max (Walker et al. 2000).

Blood Lactate Kinetics

Along with the previously mentioned metabolic changes, blood lactate kinetics may also be affected by a tapered training interval, during both maximal and submaximal exercise.

Maximal Exercise

Peak *blood lactate concentration* after maximal exercise can increase as a result of tapering. This change, which could be related to an increased posttaper muscle glycogen concentration by a mass-action effect (Houmard et al. 1994), might underpin enhanced maximal performance capabilities (Bonifazi et al. 2000, Mujika et al. 2000, Mujika et al. 2002a). A number of studies on this matter are discussed next. Table 2.3 summarizes these studies along with a number of others.

In a study on middle-distance runners, percentage 800 m competitive performance change during the taper positively correlated with changes in peak postrace blood lactate concentration (r = .87) (Mujika et al. 2000). In a subsequent study by the same group, postrace peak blood lactate increased by 7.6%, with peak lactate correlating highly with running performance (Mujika et al. 2002a). Statistically significant relationships between increases in peak postrace blood lactate levels and competition performance enhancement (r = .63) were seen in international-level swimmers during two consecutive seasons (Bonifazi et al. 2000). Male cyclists' peak blood lactate concentration increased by 78% after 14 days of a step taper, concomitant with a 7.2% improvement in an 8.5 km outdoor time trial and a 10.3% increase in peak power output (Jeukendrup et al. 1992). Competitive high school swimmers increased their peak blood lactate concentration by 20% after taper programs that induced time-trial performance gains of 4% to 8% (D'Acquisto et al. 1992).

Shepley and colleagues (1992) reported modest peak blood lactate increases after high-intensity (7.2%) and low-intensity (9.8%) tapers, whereas a laboratory-based performance measure (time to exhaustion at 1,500 m run pace) improved by 22% and 6%. Van Handel and colleagues (1988) also showed modest changes in peak lactate concentrations in collegiate swimmers preparing for national championships (6.9-7.5 mM). A similar trend (peak blood lactate concentration increased from 14.4 to 15.8 mM) was observed in elite junior rowers after a 1-week taper (Steinacker et al. 2000).

In contrast, no changes in peak blood lactate values or performance were observed in a study in which rowers performed maximal 500 m indoor rowing tests before and after a 1-week taper subsequent to 3 weeks of overload training. The author of the study concluded that the 25% decrease in training volume during the taper was insufficient or the length of the taper was too short for positive regenerative adaptations and improved performance to occur (Smith 2000). Papoti and colleagues (2007) also reported statistically unchanged peak blood lactate values in 16-year-old swimmers tapering for 11 days (peak blood lactate 6.8 and 7.2 mM before and after the taper, respectively). These authors speculated that the taper promoted an *intramuscular creatine phosphate supercompensation* (Papoti et al. 2007).

Table 2.3 Effects of the Taper on Blood Lactate Concentration (H[La])

Study (year)	Athletes	Taper duration, days	Peak H[La]	Submaximal H[La]	Performance measure	Performance outcome, %
Costill et al. (1985)	Swimmers	14	NR	↓	46-1,509 m competition	2.2-4.6 impr
Van Handel et al. (1988)	Swimmers	20	↑ slightly	↑ slightly	NR	NR
D'Acquisto et al. (1992)	Swimmers	14-28	↑	↓	100 m, 400 m time trial	4.0-8.0 impr
Jeukendrup et al. (1992)	Cyclists	14	↑	NR	8.5 km outdoor time trial	7.2 impr
Johns et al. (1992)	Swimmers	10-14	NR	↔	46-366 m competition	2.0-3.7 impr
Shepley et al. (1992)	Runners	7	↑ slightly	NR	Treadmill time to exhaustion	6-22 impr
McConell et al. (1993)	Runners	28	NR	↑	5K indoor race	1.2 decl
Flynn et al. (1994)	Runners Swimmers	21	NR NR	↔	Treadmill time to exhaustion 23 m, 366 m time trial	↔ ≈3 impr
Houmard et al. (1994)	Runners	7	NR	↔	5K treadmill time trial	2.8 impr
Stone et al. (1996)	Weightlifters	7-28	NR	↔	Competition	8.0-17.5 kg impr
Kenitzer (1998)	Swimmers	14-28	NR	↓ 14 days, ↑ 21-28 days	4 × 91 m submaximal set	≈4 impr
Bonifazi et al. (2000)	Swimmers	14-21	↑	NR	100-400 m competition	1.5-2.1 impr
Mujika et al. (2000)	Runners	6	↑ slightly	NR	800 m competition	↔
Smith (2000)	Rowers	7	↔	NR	500 m simulated time trial	↔
Steinacker et al. (2000)	Rowers	7	↑ slightly	NR	2,000 m time trial-competition	6.3 impr
Rietjens et al. (2001)	Cyclists	21	NR	↔	Incremental maximal test	↔
Mujika et al. (2002a)	Runners	6	↑	NR	800 m competition	0.4-1.9 impr
Papoti et al. (2007)	Swimmers	11	↑ slightly	NR	200 m time trial	1.6 impr

decl = decline; impr = improvement; NR = not reported; ↓ indicates decreased; ↑ indicates increased; ↔ indicates unchanged.

Adapted, by permission, from I. Mujika, S. Padilla, D. Pyne, et al., 2004, "Physiological changes associated with the pre-event taper in athletes," *Sports Medicine* 34: 903.

How Blood Lactate Levels Indicate Athlete Adaptations

When exercising muscles use carbohydrates, they produce lactic acid, which immediately splits into lactate and hydrogen ions. Increasing exercise intensity translates into higher lactate production. Lactate diffuses from the muscles to the blood, and blood lactate is used by sport scientists as an index of the physiological demands of a given exercise bout. Fatigued and carbohydrate-depleted muscles struggle to produce power and lactate (as during times of heavy training), whereas well-recovered muscles can work fast and push hard, producing lots of lactate (this should happen during the taper). Measuring blood lactate after intense exercise can therefore provide valuable information about the adaptations of an athlete to the taper.

Submaximal Exercise

Blood lactate concentration at submaximal exercise intensity shows variable responses after the preevent taper (table 2.3). Kenitzer (1998) described a decrease in blood lactate concentration at 80% of maximal HR during the first 2 weeks of a taper in female swimmers but a subsequent increase during weeks 3 and 4, leading to the tentative conclusion that 2 weeks was the optimum taper duration. In contrast, D'Acquisto and colleagues (1992) observed reduced blood lactate values during submaximal swimming in high school athletes tapering for either 2 weeks (15-26% decline) or 4 weeks (26-33% decline). These results are consistent with those of Costill and colleagues (1985), who described a 13% reduction in submaximal lactate, in parallel with a mean 3.1% swimming performance improvement in competition after a 2-week taper. Similarly, an 8.0% higher posttaper power output at a blood lactate concentration of 4 mM has been reported in rowers (Steinacker et al. 2000).

In contrast to investigators who reported higher blood lactates, Flynn and colleagues (1994) failed to observe any change in blood lactate concentration in collegiate athletes running at 75% $\dot{V}O_2$max or swimming at 90% $\dot{V}O_2$max. Unchanged submaximal blood lactate concentrations were also observed in runners performing either a run or a cycle taper for 7 days (Houmard et al. 1994), in collegiate swimmers tapering for either 10 or 14 days before a major competition (Johns et al. 1992), in cyclists performing a step taper for 21 days (Rietjens et al. 2001), and in elite weightlifters tapering for 1 or 4 weeks (Stone et al. 1996). Van Handel and colleagues (1988) reported a subtle shift of the *blood lactate–swimming velocity curve* back to the left after taper, whereas *lactate recovery curves* were unaffected. McConell and colleagues (1993), on the other hand, reported a higher blood lactate at 95% $\dot{V}O_2$max in runners after a 4-week step taper. Inconsistent findings may be related to the duration and the type of training performed during the taper.

Blood Lactate and the Taper   AT A GLANCE

The findings reported here and other similar results (Chatard et al. 1988, Lacour et al. 1990, Telford et al. 1988) on blood lactate concentration after maximal exercise support the contention that postcompetition peak blood lactate values may be a useful index of anaerobic capacity (Lacour et al. 1990, Mujika et al. 2000) and a sensitive marker of tapering-induced physiological changes (Mujika et al. 2000, Mujika et al. 2002a). Indeed, higher peak blood lactate concentrations after a taper have been related to enhanced maximal performance capabilities in swimming, running, cycling, and rowing. In contrast, unchanged or reduced blood lactate concentrations at submaximal exercise intensities should be expected after an efficient taper.

Blood Ammonia

When the rate of muscular *adenosine triphosphate (ATP) hydrolysis* exceeds the rate of *adenosine diphosphate (ADP) rephosphorylation* through either oxidative or nonoxidative processes, ATP is resynthesized via the myokinase reaction. This process results in the formation of adenosine monophosphate (AMP), which is then deaminated leading to the production of inosine monophosphate and ammonia (Lowenstein 1990, Sahlin and Broberg 1990, Tullson and Terjung 1991). Postexercise blood ammonia levels are used as a marker of exercise-induced adenine nucleotide degradation and to monitor training stress and overtraining (Warren et al. 1992).

Blood Ammonia as a Marker

When high-intensity exercise is performed, there may be an imbalance between the energy supply and the demand for ATP, eventually leading to the formation of ammonia. Ammonia is also produced when muscle protein reserves are broken down and oxidized during exercise, especially when carbohydrate availability is limited. Blood ammonia concentration can thus be used as a marker of the muscles' ability to produce ATP from ADP and also as evidence of increased protein degradation during times of increased training stress.

Mujika and colleagues (1996d) observed stable resting ammonia concentrations during a 4-week taper, but mean posttaper values (34.1 μmol/L) returned to baseline (32.8 μmol/L) after they had been elevated by 12 weeks of intensive training (65.6 μmol/L). Plasma ammonia values, however, did not reflect changes in competition performance. Elite weightlifters' resting and postexercise blood ammonia concentrations were not changed by 1 or 4 weeks of taper, during which performance in competition increased by 8 and 17.5 kg (Stone et al. 1996). Similarly, blood ammonia was not affected by 1 week of taper in elite rowers performing a maximal 500 m rowing ergometer time trial (Smith 2000). Given the potential influence of the taper on exercise metabolism and preexercise energy substrate availability, further research on blood ammonia changes associated with different tapering strategies is warranted, especially in high-intensity, short-duration sporting events.

Muscle Glycogen

Muscle glycogen concentration has been shown to increase progressively during tapering. Neary and colleagues (1992) showed an increase in muscle glycogen concentration of 17% after a 4-day taper and 25% after an 8-day taper. The same group of investigators compared the change in muscle glycogen concentration after two different 7-day tapers in male cyclists: one in which training intensity was maintained at around 85% to 90% of maximal HR and training duration progressively reduced from 60 to 20 min and a second where the duration remained constant at 60 min but intensity declined progressively from 85% to 55% of maximal HR. Muscle glycogen concentration increased by 34% and simulated 40 km time trial performance by 4.3% during the shorter, more intense taper, whereas muscle glycogen concentration increased by 29% and performance by 2.2% in the longer, less intense taper (Neary et al. 2003b). A similar high-intensity, low-duration taper brought about a 15% increase in muscle glycogen (figure 2.6) and 22% in running performance on a treadmill, whereas no change in glycogen and a moderate 6% performance gain were observed with a low-intensity, moderate-duration taper (Shepley et al. 1992).

Six days of tapering during the *luteal phase* of *eumenorrheic* athletes' menstrual cycles resulted in a 13% higher muscle glycogen level when a high-carbohydrate diet (78% carbohydrate) was consumed during the final 3 to 4 days of the taper, in comparison with a moderate-carbohydrate diet (48% carbohydrate). The women were able to supercompensate glycogen during the taper but not to the same magnitude as generally reported in men. Concomitantly, cycling time to exhaustion at 80% to 82% $\dot{V}O_2$max increased by 8% (Walker et al. 2000).

Mineral Metabolism

Mineral metabolism (basal plasma and 24 hr urinary calcium, magnesium, iron, zinc, and copper) was studied in nine male cyclists during 6 weeks of volume training, 18 days of high-intensity interval training, and 10 days of a reduced frequency taper. Urinary calcium decreased by 11.4% and plasma calcium concentration increased by 5.1% with the taper, but the metabolism of all other studied minerals remained unchanged (Dressendorfer et al. 2002b). The apparent rebound in renal calcium might have resulted either from decreased urinary calcium filtration associated with lower ionized plasma calcium or from increased plasma parathyroid hormone levels. Irrespective of the regulating mechanism, reducing the frequency of interval training during the taper appeared to trigger compensatory calcium retention (Dressendorfer et al. 2002b).

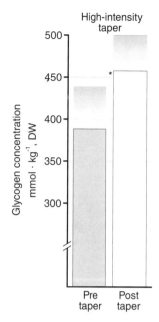

Figure 2.6 Resting muscle glycogen concentration for vastus lateralis before and after a high-intensity taper ($N = 8$). Values are mean ± SD and are expressed per unit dry weight. *Significant pre- to post-taper differences, $p < .05$.

Adapted, by permission, from B. Shepley, J.D. MacDougall, N. Cipriano, et al., 1992, "Physiological effects of tapering in highly trained athletes. *Journal of Applied Physiology* 72(Feb): 709. Permission conveyed through Copyright Clearance Center, Inc.

Energy for Exercise and the Taper | AT A GLANCE

Reduced training loads will affect daily energy expenditure, and athletes should pay special attention to their energy intake during the taper to avoid energy imbalance and undesirable changes in body composition. Substrate contribution to power production during exercise usually remains stable during the taper, but an elevated contribution from carbohydrates could take place attributable to increased muscle glycogen concentration. Increased peak blood lactate concentrations following the taper are often related to improved maximal performance capabilities in cyclic sports. During submaximal-intensity exercise, blood lactate concentrations may be unchanged or reduced after the taper. Blood ammonia concentrations have been measured to assess changes in adenine nucleotide degradation during the taper, but results have not been definitive.

Chapter Summary

Physiological mechanisms underlying the performance gains associated with the taper are slowly being unveiled by sport scientists. Increases in maximal oxygen uptake during the taper may contribute to posttaper performance gains, but improved performances can also take place without changes in maximal oxygen uptake. A reduced oxygen cost of exercise after the taper (i.e., a better economy of movement) can also contribute to performances gains, but this change is more likely to be observed in athletes of a lower caliber. Cardiac function and dimensions and ventilatory function do not generally change as a result of the taper, and any observed variation could be a reflection of the athlete's delayed adaptation to prior training rather than to the taper itself.

Hematological changes strongly suggest that the training reductions usually applied during the taper facilitate a positive balance between erythropoiesis (i.e., red cell production) and hemolysis (i.e., red cell destruction), which contributes to taper-induced performance improvements. However, increased red cell production could compromise the iron status of the athlete because erythropoiesis requires significant iron.

Reduced training loads will also affect daily energy expenditure, and athletes should pay special attention to their energy intake during the taper so that energy imbalance and undesirable changes in body composition can be avoided. Carbohydrate and fat contribution to power production during exercise usually remains stable during the taper, but an elevated contribution from carbohydrates could take place attributable to increased muscle glycogen concentration.

Higher peak blood lactate concentrations after a taper have been related to enhanced maximal performance capabilities in various modes of locomotion, including swimming, running, cycling, and rowing. To the contrary, the taper usually results in unchanged or reduced blood lactate concentrations at submaximal exercise intensities. Blood ammonia concentrations have been measured to assess changes in adenine nucleotide degradation and protein breakdown during the taper, but results have been inconclusive.

chapter 3

Taper-Associated Biochemical, Hormonal, Neuromuscular, and Immunological Changes

In addition to the cardiorespiratory and metabolic changes that often take place when athletes are tapering for major competition, which were analyzed in the previous chapter, other physiological systems adapt to the taper program. As a result, biochemical, hormonal, neuromuscular, and immunological changes may take place during this time. These changes contribute to an athlete's enhanced performance capabilities and can thus be indicative of an athlete's preparedness to perform.

Biochemical Adaptations

Few biochemical parameters have been shown to exhibit marked changes during the preevent taper, limiting their utility as markers of physiological recovery and increased performance capacity. One of the most widely used and studied biochemical markers of training stress is the concentration of creatine kinase in the blood.

Creatine Kinase

Blood levels of creatine kinase (CK) have been used as an index of training-induced physiological stress. Creatine kinase is a muscle enzyme that occasionally increases in the blood following strenuous or eccentric exercise, most probably as a result of altered permeability of tissue cell membranes. Factors that influence the degree of CK efflux into the blood include exercise duration and intensity, exercise mode, and fitness level of the individual (Millard et al. 1985).

Creatine Kinase

Creatine kinase is an important enzyme in muscle energy production that is generally confined to the inside of muscle cells. The presence of elevated values of CK in the blood suggests that muscle cell membranes have suffered some kind of damage, allowing CK to escape the cells. Following intensive training, the blood concentration of CK is often remarkably increased, suggesting increased levels of muscle tissue breakdown.

Table 3.1   Effects of the Taper on Blood Creatine Kinase (CK) Concentration

Study (year)	Athletes	Taper duration, days	Blood CK concentration	Performance measure	Performance outcome, %
Burke et al. (1982b)	Swimmers	28	↓	NR	NR
Millard et al. (1985)	Swimmers	28	↓	NR	NR
Yamamoto et al. (1988)	Swimmers	14-26	↓	NR	NR
Houmard et al. (1990b)	Runners	21	↓	5K indoor race	↔
Costill et al. (1991)	Swimmers	14-21	↓	Competition	≈3.2 impr
Flynn et al. (1994)	Runners Swimmers	21	↔ ↓	Treadmill time to exhaustion 23 m, 366 m time trial	↔ ≈3 impr
Mujika et al. (1996d)	Swimmers	28	↓	100-200 m competition	0.4-4.9 impr
Hooper et al. (1999)	Swimmers	14	↑ slightly	100 m time trial	↔
Child et al. (2000)	Runners	7	↓	Simulated half-marathon	↔
Mujika et al. (2000)	Runners	6	↔	800 m competition	↔
Mujika et al. (2002a)	Runners	6	↔	800 m competition	0.4-1.9 impr
Coutts et al. (2007a)	Triathletes	14	↔	3 km run time trial	3.9 impr

impr = improvement; NR = not reported; ↓ indicates decreased; ↑ indicates increased; ↔ indicates unchanged.

Adapted, by permission, from I. Mujika, S. Padilla, D. Pyne, et al., 2004, "Physiological changes associated with the pre-event taper in athletes," *Sports Medicine* 34: 905.

Various studies have shown decreases in CK levels during the taper (table 3.1). Studying 10 male and 10 female collegiate swimmers before and after a 4-week taper, Millard et al. (1985) noted a 70% lower posttraining and a 30% lower resting serum CK after the taper in the males and 28% and 7% lower values in the females. Absolute posttaper CK values were not different between genders and fell to their lowest levels of the season during the taper (figure 3.1). These results suggested that CK levels reflect training volume rather than intensity. Yamamoto and colleagues (1988) also observed decreased CK levels after swimming tapers correlating with the daily workout volume during the taper.

Flynn and colleagues (1994) described a CK reduction of 38% during 3 weeks of taper. Mujika and colleagues (1996d) also reported a 43% decline in plasma CK during a 4-week taper, but this did not correlate with swimming performance improvements, which ranged from 0.4% to 4.9%. Costill and colleagues (1991) measured 28% lower CK values after 2 to 3 weeks of taper, which resulted in an average performance improvement of 3.2%. Burke and colleagues (1982b) also observed a decline in CK levels after the taper in swimmers, but their values remained in the high–normal range.

In a group of male distance runners performing simulated half-marathons before and after either a 7-day taper or normal training, lower serum CK values were measured after the second half-marathon in the tapered athletes (Child et al. 2000). This attenuated release of CK into the bloodstream was unlikely to be a consequence of lower muscle force generation, reduced oxidative stress, or increased extracellular antioxidant protection. It was speculated that the taper may have facilitated muscle recovery, such that CK returned to resting values prior to the second half-marathon and resistance to exercise-induced injury was similar to that observed before the taper. Despite evidence that the taper reduced muscle damage, half-marathon performance did not improve (Child et al. 2000). A 3-week step taper also resulted in decreased serum CK in well-trained runners, but performance did not change (Houmard et al. 1990b).

In contrast to studies showing lower CK values after the taper, Hooper and colleagues (1999) measured a statistically nonsignificant 17% increase in plasma CK during a 2-week taper, with very large interindividual variation among swimmers. These authors argued that plasma CK is not a reliable marker of training stress and reflects an acute response to a single exercise session rather than an athlete's homeostatic status. They also suggested that the large interindividual variation could indicate large differences in athletes' physiological responses to the taper. Very large interindividual variations with statistically unchanged means were also observed in triathletes after a 2-week taper (Coutts et al. 2007b) and in middle-distance runners after 6-day tapers (Mujika et al. 2000, Mujika et al. 2002a). Nevertheless, plasma CK levels increased with increasing low-intensity continuous training distance during the taper, which led the authors to suggest that

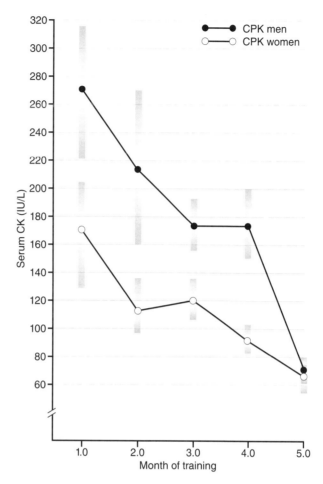

Figure 3.1 Serum creatine kinase values observed postexercise during months 1 through 5 of the swimming season. Symbols indicate values ± SEM.

Reprinted, by permission, from M. Millard, C. Zauner, R. Cade, et al., 1995, "Serum CPK levels in male and female world class swimmers during a season of training," *Journal of Swimming Research* 1: 12-16.

middle-distance runners could limit the exercise-induced skeletal muscle damage before competition by reducing continuous running volume during the taper (Mujika et al. 2000).

The published literature appears to suggest that plasma CK values could be of some interest to assess recovery from acute training stress and muscle damage during the taper, but the validity of this parameter as a marker of an individual athlete's performance capabilities seems limited.

Other Biochemical Markers

Yamamoto and colleagues (1988) measured serum glutamic oxalacetic transaminase and glutamic pyruvic transaminase (GPT) during two separate tapers in male swimmers, with declines in GPT in the initial 4 to 9 days of each of the tapers. The authors speculated that this decrease may reflect an increased posttaper ATP availability, maintaining the integrity of the cellular membrane and reducing the efflux of enzymes into the bloodstream. Banister and colleagues (1992) reported marked declines in the pattern of an elevation of serum enzyme activity (including CK, lactate dehydrogenase, and aspartate aminotransferase) during 32 days of taper subsequent to 28 days of normal training in two subjects.

Other purported biochemical markers of training stress and performance capability, including blood urea, uric acid, and creatinine, were unchanged during 6-day tapers in middle-distance runners (Mujika et al. 2000, Mujika et al. 2002a) and a 14-day taper in triathletes (Coutts et al. 2007b). Serum uric acid did not change in distance runners during a 7-day taper (Child et al. 2000). Costill and colleagues (1985) did not detect any taper-induced changes in their swimmers' blood pH, partial pressure of carbon dioxide (PCO_2), partial pressure of oxygen (PO_2), bicarbonate (HCO_3), and base excess after a 200-yard (183 m) submaximal swim at 90% of the season's best performance.

AT A GLANCE   **Biochemical Markers of Training Stress**

Creatine kinase is one of the most widely used blood parameters purported to provide an indication of training-induced physiological stress. Reduced blood CK concentrations after a training taper suggest recovery from training stress and muscle damage, but these reductions do not necessarily correlate with improved performance. Other biochemical markers of training stress and performance capacity are largely unaffected by the taper.

Hormonal Adaptations

Strenuous physical exercise is known to result in short- and long-term alterations of the endocrine system (Bunt 1986; Galbo 1986; Viru 1992). Because of their responsiveness to training-induced physiological stress, various hormones including testosterone, cortisol, catecholamines, growth hormone, and others are often used as markers to monitor training stress, evaluate training responses, and predict performance capacity. These hormonal markers of training stress should therefore reflect the variations in the training load that take place during different phases of a competitive season, such as during a taper, and hormonal changes should be related with changes in competition performance.

Testosterone, Cortisol, and the Testosterone–Cortisol Ratio

The plasma levels of testosterone (T) and cortisol (C) could represent anabolic and catabolic tissue activities, respectively. Although the T:C ratio has been suggested as a marker of training stress (Adlercreutz et al. 1986, Kuoppasalmi and Adlercreutz 1985), the available data in the literature concerning androgen and C responses to tapering in athletes are inconclusive (table 3.2).

In a study on collegiate runners and swimmers, Flynn and colleagues (1994) observed no change in the runners' total testosterone (TT), free testosterone (FT), and TT:C or FT:C ratios during 3 weeks of taper. Within the same study, on the other hand, a group of swimmers' TT and FT returned toward baseline during the taper, after showing blunted values throughout the intensive training phases of the season. No changes were noticed in the TT:C and FT:C ratios. Interestingly, changes in TT and FT during training and taper, but not in TT:C or FT:C ratios, paralleled changes in performance during criterion swims (Flynn et al. 1994). Male cyclists' serum T increased by 5.3%, and 24 hr urinary C decreased marginally by 4.6% during 10 days of reduced training, in the face of a 1.2% improvement in performance. In addition, the ratio of serum T to urinary C ratio was unaltered by the taper (Dressendorfer et al. 2002a).

No changes in resting blood levels of T or C or in TT:C ratios were found in runners performing a 3-week step taper (Houmard et al. 1990b), nor were changes in these variables and sex-hormone binding globulin (SHBG) found in elite weightlifters after either 1 or 4 weeks of taper, during which competition performance improved by 8.0

Table 3.2 Effects of the Taper on Blood Testosterone and Cortisol Concentrations

Study (year)	Athletes	Taper duration, days	Blood testosterone concentration	Blood cortisol concentration	Testosterone–cortisol ratio	Performance measure	Performance outcome, %
Houmard et al. (1990b)	Runners	21	↔	↔	↔	5K indoor race	↔
Costill et al. (1991)	Swimmers	14-21	↑	↓	NR	Competition	≈3.2 impr
Tanaka et al. (1993)	Swimmers	14	↔	↔	↔	NR	NR
Flynn et al. (1994)	Runners Swimmers	21	↔ ↑	↔ ↔	↔ ↔	Treadmill time to exhaustion 23 m, 366 m time trial	↔ ≈3 impr
Mujika et al. (1996d)	Swimmers	28	↔	↔	↔	100-200 m competition	0.4-4.9 impr
Stone et al. (1996)	Weightlifters	7-28	↔	↔	↔	Competition	8.0-17.5 kg impr
Bonifazi et al. (2000)	Swimmers	14-21	NR	↓	NR	100-400 m competition	1.5-2.1 impr
Martin et al. (2000)	Cyclists	7	NR	↑	NR	Incremental maximal test	≈6 impr
Mujika et al. (2000)	Runners	6	↔	↔	↔	800 m competition	↔
Steinacker et al. (2000)	Rowers	7	↑ slightly	↑ slightly	NR	2,000 m time trial–competition	6.3 impr
Dressendorfer et al. (2002a)	Cyclists	10	↑ slightly	NR	NR	20 km simulated time trial	1.2 impr
Mujika et al. (2002a)	Runners	6	↑	↔	↔	800 m competition	0.4-1.9 impr
Maestu et al. (2003)	Rowers	14	↓	↔	NR	2000 m ergometer time trial	Unchanged
Coutts et al. (2007a)	Triathletes	14	↔	↓	↑	3 km run time trial	3 impr
Izquierdo et al. (2007)	Strength trained	28	↔	↔	NR	1RM strength	2.0-3.0 impr

impr = improvement; NR = not reported; ↓ indicates decreased; ↑ indicates increased; ↔ indicates unchanged.

Adapted, by permission, from I. Mujika, S. Padilla, D. Pyne, et al., 2004, "Physiological changes associated with the pre-event taper in athletes," *Sports Medicine* 34: 907.

and 17.5 kg (Stone et al. 1996). Strength-trained athletes tapering for 4 weeks after 16 weeks of heavy resistance training did not change their resting serum TT, FT, and C concentrations either (Izquierdo et al. 2007).

Mujika and colleagues have assessed the effects of tapering on selected hormones in swimmers and middle-distance runners (Mujika et al. 1996d, Mujika et al. 2000, Mujika et al. 2002a). In a study on swimmers, plasma TT, non–SHBG-bound testosterone (NSBT; which is the sum of FT and albumin-bound T representing the biologically active fraction of T) (Cumming and Wall 1985, Manni et al. 1985), C, TT:C, and NSBT:C remained stable during a 4-week taper subsequent to 8 weeks of intensive training, despite large variations in training volume (Mujika et al. 1996d). Nevertheless, the 4 weeks of taper resulted in a 2.3% improvement in competition performance, and percentage variations in swimming performance during the taper correlated with changes in the TT:C ($r = .81$) (see figure 1.3) and NSBT:C ($r = .76$) ratios and with changes in NSBT concentration ($r = .71$) (Mujika et al. 1996d). In contrast, Coutts and colleagues (2007b) observed decreased C and increased FT:C in well-trained triathletes tapering for 2 weeks, but these changes did not correlate with observed gains in performance.

In an initial 6-day taper study on 800 m runners, TT, FT, C, TT:C, and FT:C remained stable as a result of the taper, and individual changes in these markers did not parallel changes in performance during the taper. However, TT correlated inversely with low-intensity continuous training distance during the taper and positively with high-intensity interval training distance, suggesting that low-intensity continuous training may hinder anabolic processes stimulated by testosterone during the taper, whereas they could be facilitated by high-intensity interval training (Mujika et al. 2000) (figure 3.2). A subsequent study on 800 m runners who also tapered for 6 days showed increased TT values after the taper, attributed to the elimination of low-intensity continuous training from the main part of the training sessions during the taper. The mechanism for increased TT posttaper is thought to relate to enhanced pituitary response to the preceding time of intense training, bringing about a positive influence on *androgenic–anabolic activity* during the subsequent taper, characterized by reduced levels of physiological stress (Busso et al. 1992, Mujika et al. 1996d, Mujika et al. 2002a).

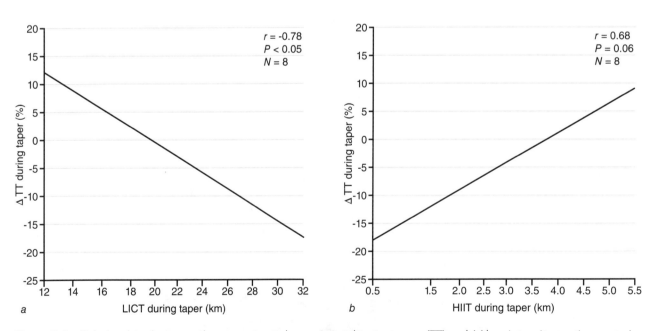

Figure 3.2   Relationships between the percentage change in total testosterone (TT) and *(a)* low-intensity continuous training (LICT) distance during the taper and *(b)* high-intensity interval training (HIIT) distance during the taper.

Adapted, by permission, from I. Mujika, A. Goya, S. Padilla, et al., 2000, "Physiological responses to a 6-day taper in middle-distance runners: Influence of training intensity and volume," *Medicine & Science in Sports & Exercise* 32: 511-517.

In view of the relationship between *luteinizing hormone (LH)* and T, regulated by the hypothalamic–pituitary–testicular axis, and given that the C secretion is in part controlled by a common regulatory pathway (i.e., the hypothalamic–pituitary–adrenocortical axis), Mujika and colleagues (2004) suggested that future investigations should assess possible changes in the adrenocorticotropic hormone (ACTH) concentration during the taper. This was recently done by Coutts and colleagues (2007b) in triathletes undergoing 4 weeks of intensified training followed by 2 weeks of tapering. The observed trends in ACTH concentrations were indeed suggestive of an enhanced pituitary response during the taper after the intensified training, as mentioned previously.

Changes in resting C concentration during the taper have been proposed as a means of monitoring performance capacity in athletes (table 3.2). For instance, Mujika and colleagues (1996b) found slight reductions in resting cortisol concentrations in swimmers who responded to a 4-week taper by improving their performance by more than 2% but a significant increase in cortisol in swimmers less respondent to the same taper program, suggesting a relationship between resting cortisol levels and the performance response to the taper (figure 3.3). Collegiate swimmers' resting C values declined by 23% to 30%, their T concentration increased by 22% during the first taper, and the athletes' competition performance improved by an average of 3.2% in two different 2- to 3-week tapers within a season (Costill et al. 1991). On the other hand, no changes in TT, C, or the TT:C ratio were observed during 6 weeks of progressive increase and 2 weeks of gradual decrease in training volume in well-trained swimmers (Tanaka et al. 1993).

A follow-up investigation on elite swimmers over two seasons showed that the 1.5% to 2.1% performance improvements during the tapers before the major competitions of each season were positively related to the corresponding 22% to 49% increases in postcompetition peak lactate concentrations but negatively related ($r = -.66$) to the 19% to 29% change in resting precompetition plasma C concentration (Bonifazi et al. 2000) (figure 3.4). The conclusion of this study was that a low C concentration was a prerequisite for improved performance in events that rely largely on the contribution of anaerobic metabolism to total energy supply (Bonifazi et al. 2000). In keeping with this conclusion, Mujika and colleagues (2002a) observed strong correlations between changes in peak blood lactate concentration after an 800 m running race during a 6-day taper and changes in serum C ($r = -.75$) and the TT:C and FT:C ratios ($r = .82$) (figure 3.5). Collectively, these findings indicate that a hormonal milieu propitious to anabolic processes is necessary for optimum function of the glycolytic power system and performance in middle-distance events.

Elite junior rowers showing clear signs of overreaching and hypothalamic down-regulation during a high-load training phase reportedly recovered during a subsequent week of taper.

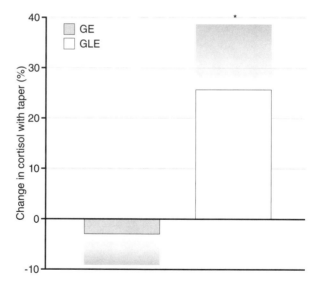

Figure 3.3 Group comparison for the percentage of variation in blood cortisol with the taper. GE (group efficient) and GLE (group less efficient): swimmers improving their competition performance with the taper by more than 2% and less than 2%, respectively. *$p < .05$.

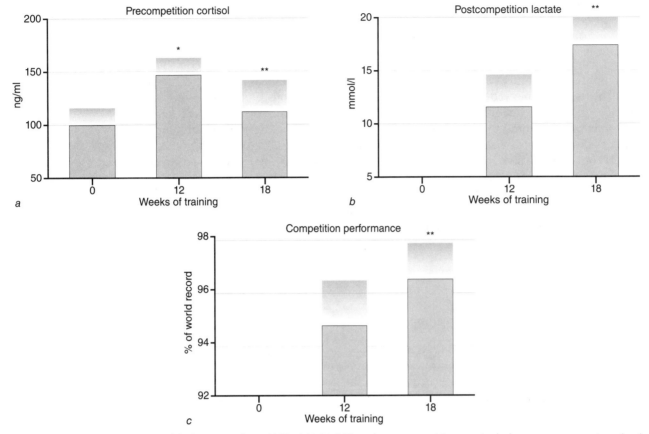

Figure 3.4 1997 season (World Championships, WC): *(a)* initial (0) and precompetition cortisol plasma concentrations (ng/ml); *(b)* postcompetition blood lactate concentration (mmol/l) and *(c)* competition performances expressed as a percentage of the world record in the preparatory competition (PWC; after 12 weeks of training) and WC (after tapering, 18 weeks). Data are given as mean and standard deviation. *T* test for paired data: cortisol, 12 weeks significantly different from initial (**p* = .003), 18 weeks significantly different from 12 weeks (***p* = .029); lactate, 18 weeks significantly different from 12 weeks (***p* < .001); performance, 18 weeks significantly different from 12 weeks (***p* < .001).

With kind permission from Springer Science + Business Media: *European Journal of Applied Physiology,* "Preparatory versus main competitions: Differences in performances, lactate responses and pre-competition plasma cortisol concentrations in elite male swimmers," vol. 82, 2000, p. 370, M. Bonifazi, F. Sardella, and C. Luppo, fig. 1.

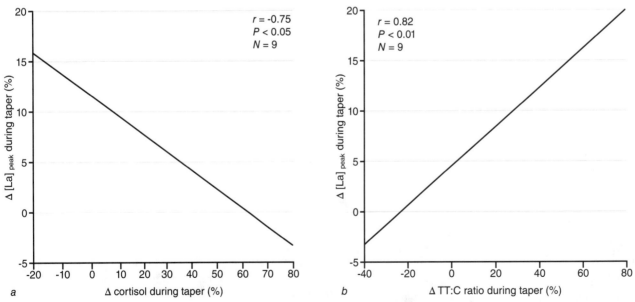

Figure 3.5 Relationships between the percentage change in peak blood lactate concentration ([La]$_{peak}$) during the taper and *(a)* percentage change in serum cortisol concentration during the taper and *(b)* percentage change in the total testosterone–cortisol ratio (TT:C) during the taper.

Adapted, by permission, from I. Mujika, A. Goya, E. Ruiz, et al., 2002, "Physiological and performance responses to a 6-day taper in middle-distance runners: Influence of training frequency," *International Journal of Sports Medicine* 23: 367-373.

Time trial (2,000 m) performance improved by 6.3%, whereas peripheral and central steroid hormone concentrations increased by about 10% (Steinacker et al. 2000) (figure 3.6). These changes were accompanied by positive changes in the Recovery–Stress Questionnaire for Athletes, which led the authors to suggest that the hypothalamus plays an important role in integrating different kinds of stress influences and responds by means of the endocrine system (e.g., pituitary–adrenocortical and pituitary–gonadal axes), the autonomic nervous system, and behavior (Steinacker et al. 2000).

Female collegiate swimmers' salivary C levels have been reported to return to baseline values after 4 weeks of taper, consisting of a progressive 63% reduction in training volume (O'Connor et al. 1989). A similar finding was reported in male swimmers, whose salivary C decreased marginally by 4.8% during the taper, to attain the lowest values of the entire training season (Tharp and Barnes 1990). Several investigations, however, have reported unchanged or slightly increased C concentrations as a result of a taper in swimmers (Mujika et al. 1996d, Mujika et al. 2002a), cyclists (Martin et al. 2000), and rowers (Steinacker et al. 2000). Plasma C concentrations are subject to different kinds of physiological and psychological stressors (McCarthy and Dale 1988, Stein et al. 1985), which could explain conflicting findings. The physical stress produced by the pretaper intensive training could be replaced in the posttaper condition by the psychological stress associated with the oncoming major competition (Mujika et al. 1996c).

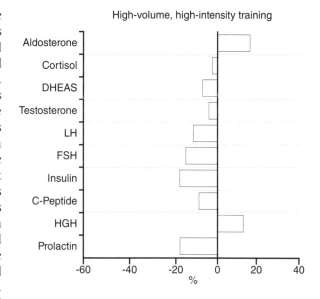

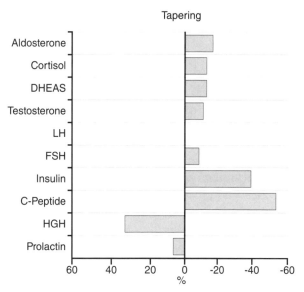

Figure 3.6   Change in resting hormone concentrations at the beginning of the training camp, at the end of phase 2 with high-volume, high-intensity training, and at the end of the first tapering segment (phase 4): *aldosterone*, cortisol, dehydroepiandrosterone (DHEAS), free testosterone, luteinizing hormone (LH), follicle-stimulating hormone (FSH), insulin, *C-peptide*, human growth hormone (hGH), and *prolactin*. Percentage change of median values.

Reprinted, by permission, from J.M. Steinacker, W. Lormes, M. Kellmann, et al., 2000, "Training of junior rowers before world championships. Effects on performance, mood state and selected hormonal and metabolic responses," *Journal of Sports Medicine and Physical Fitness* 40(4): 332.

Atlaoui and colleagues (2004) speculated that explanations for discrepancies among studies include C binding with corticosteroid binding globulin, varying densities of receptors in target tissues, and prereceptor metabolism of C by the tissue-specific enzyme 11β-hydroxysteroid dehydrogenase (11β-HSD). Two isoenzymes of 11β-HSD interconvert active C and inactive cortisone (Cn) within target cells, modulating C action at an *autocrine* level in peripheral tissues. Given that the 24-hr urinary C:Cn ratio is a valid index of renal 11β-HSD activity (Best and Walker 1997), it was suggested that the C:Cn ratio could provide insight into the adaptations of the hypothalamo–pituitary–adrenal axis to training and tapering. Atlaoui and colleagues (2004) showed a decline of the C:Cn ratio during 3 weeks of taper subsequent to 4 weeks of intensive training. A large negative correlation between performance changes in competition and changes in C:Cn during the taper ($r = -.69$) suggests that this ratio could be a useful tool for monitoring training and tapering-induced adaptations.

Catecholamines

Plasma and urinary catecholamine concentrations have been measured as a means to monitor training stress and identify overreaching or overtraining in athletes (Coutts et al. 2007a, Hooper et al. 1993, Hooper et al. 1995, Kindermann 1988, Kirwan et al. 1988, Lehmann et al. 1991, Lehmann et al. 1992, Mackinnon et al. 1997). Hooper and colleagues reported that plasma noradrenaline was higher during the taper in swimmers who were overtrained and failed to improve their performance (Hooper et al. 1993). In another study by the same group, a small change in time-trial swimming performance during the taper was predicted by changes in plasma noradrenaline concentration, which accounted for 82% of the variance. The authors concluded that the change in plasma noradrenaline concentration could be a useful marker for monitoring recovery associated with the taper, but they acknowledged that the role of catecholamines in the recovery phase after intense training is not well established and that the expense and complexity associated with catecholamine measurements preclude its use in routine screening of athletes during the taper (Hooper et al. 1999). Other investigators have also measured plasma adrenaline, noradrenaline, and dopamine concentrations before and after taper in competitive swimmers (Mujika et al. 1996d). The only noticeable change in this study was a statistically nonsignificant 22% decrease in plasma noradrenaline during the taper, but in contrast with the previously mentioned investigation, this change did not correlate with the 2.3% gains in competition swimming performance, which ranged between 0.4% and 4.9% (Mujika et al. 1996d). More recently, Coutts and colleagues (2007b) failed to observe significant changes in urinary adrenaline and noradrenaline concentrations during intensified training and tapering in well-trained triathletes.

Growth Hormone and Insulin-Like Growth Factor-I

Human growth hormone (hGH) levels increased 10% from baseline following 3 weeks of intensive training in rowers but decreased by 30% during the following 2-week taper phase (Steinacker et al. 2000). *Insulin-like growth factor-I (IGF-I)*, a 7.5 kDa polypeptide that plays an important role in the regulation of somatic growth, metabolism, and cellular proliferation, differentiation, and survival, has also been measured before and after taper in athletes (Koziris et al. 1999). Nine male collegiate swimmers' total serum IGF-I increased progressively by 76% above baseline during 4 months of intensive training, and these elevated values were maintained during 4 weeks of tapering. The levels of free IGF-I increased by 77% to 102% at all training measurements, including the taper. The levels of immunoreactive IGF binding protein-3 (IGFBP-3) were 30% higher after intensive training and remained elevated during tapering. In contrast, IGF binding protein-1 (IGFBP-1) declined to baseline values during tapering. Performance measures were not reported, but the authors of the study suggested that the increased total and free IGF-I and total IGFBP-3 could have played a role in the observed reductions in skinfold measurements during the season (Koziris et al. 1999). Two weeks of reduced training characterized by less intense training, no weight training, and shorter interval training in junior team handball players resulted in a 7.7% increase in IGF-I, which returned to baseline after being depressed during intensive training. This change during the taper was concomitant with gains in repeated sprint (2.1%) and vertical jump (3.2%) performance (Eliakim et al. 2002). In strength-trained athletes, 4 weeks of tapered strength training did not elicit any changes in resting serum hGH and IGF-I, but IGFBP-3 increased by 15% (Izquierdo et al. 2007), leading the authors to speculate that the elevated IGFBP-3 may have been a compensatory mechanism to accommodate the low IGF-I concentrations found during the taper to preserve IGF availability.

Other Hormones

Other hormones have been suggested as markers of training stress and overtraining (Busso et al. 1992, Carli et al. 1983, Fry et al. 1991) during tapering. To date, the relevant studies have yielded inconclusive results about the usefulness of hormonal monitoring. Thyroid-stimulating hormones, *triiodothyronine* and *thyroxine* concentrations, for instance, were not altered by a 4-week taper in male national- and international-level swimmers (Mujika et al. 1996d). Steinacker and colleagues (2000) reported increased resting insulin and C-peptide levels during a 1-week taper in rowers, suggesting a higher posttaper carbohydrate turnover capacity. This suggestion, however, was not reflected by the exercise lactate levels, which remained unchanged, and there was no indication of glycogen depletion among the participating subjects.

Maestu and colleagues (2003) evaluated fasting plasma *leptin* responses in 12 male national standard rowers who underwent 3 weeks of intensified training followed by 2 weeks of tapering, characterized by a 50% reduced training load. Leptin concentrations increased by 29% during the first week of taper and an additional 4% during the second week. The authors concluded that fasting plasma leptin is more sensitive to rapid changes in training stress than are previously used stress hormones and that leptin could be regarded as a key signal for metabolic adaptation to training stress in highly trained athletes (Maestu et al. 2003).

Hormones and the Taper

AT A GLANCE

Testosterone, cortisol, and the testosterone–cortisol ratio can provide information on the physiological stress, recovery, and performance capacity of an athlete during the taper, but performance gains also occur without concomitant changes in these parameters. The 24 hr urinary cortisol–cortisone ratio, plasma and urinary catecholamines, growth hormone, and insulin-like growth factor-I show promise as tools for monitoring training stress and tapering-induced adaptations, but further research is needed to draw solid conclusions about their validity.

Neuromuscular Adaptations

The extraordinary plasticity of skeletal muscle tissue allows it to adapt to variable levels of functional demands, neuromuscular activity, and hormonal signals and reversibly change its functional characteristics and structural composition (Gordon and Pattullo 1993, Hoppeler 1986, Kannus et al. 1992, Saltin and Gollnick 1983). A precompetition taper presumably reduces the demands placed on the neuromuscular system compared with previous phases of a training program.

Strength and Power

Increased strength and power as a result of a taper have been a common observation in different athletic activities (table 3.3). Costill and colleagues (1985) were among the first researchers to describe such gains in swimmers. These authors described an 18% improvement in swim bench power and a 25% gain in actual swim power in a group of 17 collegiate swimmers undergoing a 2-week taper. Swim power improvement correlated with a 3.1% competition performance gain ($r = .68$). The reduced training may

Table 3.3   Effects of the Taper on Muscular Strength and Power

Study (year)	Athletes	Taper duration, days	Strength and power	Performance measure	Performance outcome, %
Costill et al. (1985)	Swimmers	14	↑	46-1,509 m competition	2.2-4.6 impr
Cavanaugh and Musch (1989)	Swimmers	28	↑	46-1,509 m competition	2.0-3.8 impr
Prins et al. (1991)	Swimmers	28	↔	NR	NR
Johns et al. (1992)	Swimmers	10-14	↑	46-366 m competition	2.0-3.7 impr
Shepley et al. (1992)	Runners	7	↑	Treadmill time to exhaustion	6-22 impr
Gibala et al. (1994)	Strength trained	10	↑	Voluntary elbow flexor strength	≈7 impr
Houmard et al. (1994)	Runners	7	↔	5K treadmill time trial	2.8 impr
Martin et al. (1994)	Cyclists	14	↑	Incremental maximal test	8.0 impr
Raglin et al. (1996)	Swimmers	28-35	↑	Competition	2.0 impr
Hooper et al. (1998)	Swimmers	14	↑	100 m, 400 m time trial	↔
Hooper et al. (1999)	Swimmers	14	↔	100 m time trial	↔
Trappe et al. (2001)	Swimmers	21	↑	Competition	3.0-4.7 impr
Trinity et al. (2006)	Swimmers	21	↑	50-1500 m competition	4.5 impr
Papoti et al. (2007)	Swimmers	11	↑	200 m time trial	1.6 impr
Izquierdo et al. (2007)	Strength trained	28	↑	1RM strength	2.0-3.0 impr

impr = improvement; NR = not reported; ↑ indicates increased; ↔ indicates unchanged.

Adapted, by permission, from I. Mujika, S. Padilla, D. Pyne, et al., 2004, "Physiological changes associated with the pre-event taper in athletes," *Sports Medicine* 34: 911.

have allowed for an increase in maximal tension development through changes in the contractile mechanisms or neural controls on fiber recruitment (Costill et al. 1985). In keeping with these results, Johns and colleagues (1992) observed a 5% increase in tethered swimming power and a 2.8% improved performance in competition after 10 and 14 days of taper. National- and international-level swimmers' isolated mean arm and leg power has also been shown to increase during a 4-week taper, especially during the initial 5 to 24 s of exercise (Cavanaugh and Musch 1989). Competition performance increased by an average of 2.6% during the taper.

Raglin and colleagues (1996) also reported gains during a 4- to 5-week taper in swimming peak power (16%) and mean power (20%). In addition, they observed a 23% gain in neuromuscular function, as determined with the soleus Hoffmann reflex, an indicator of the general excitability of the α-motoneuron pool. These changes correlated with changes in power and were accompanied by a 2.0% improvement in competition velocity. The authors concluded that neurological adaptations may have a role in the performance gains that often follow the taper (Raglin et al. 1996). More recently, Trappe

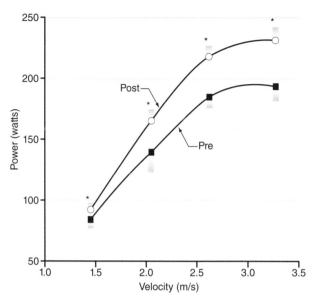

Figure 3.7 Pre- to posttaper swim bench arm power is shown at settings 0 (1.44 m/s), 3 (2.05 m/s), 6 (2.66 m/s), and 9 (3.28 m/s). *Significant p < .05 difference pre- to posttaper.

Reprinted, by permission, from S. Trappe, D. Costill, and R. Thomas, 2001, "Effect of swim taper on whole muscle and single fiber contractile properties," *Medicine & Science in Sports & Exercise* 33: 50.

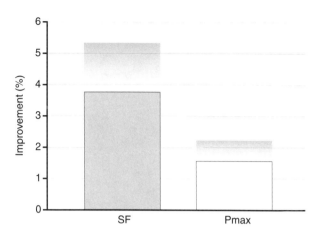

Figure 3.8 Swim force (SF) and maximal performance (Pmax) improvements after the 11-day taper.

Reprinted, by permission, from M. Papoti, L.E.B. Martins, S.A. Cunha, et al., 2007, "Effects of taper on swimming force and swimmer performance after an experimental ten-week training program," *Journal of Strength and Conditioning Research* 21(2): 540.

and colleagues (2001) noted a 7% to 20% increase in swim bench muscle power, a 13% increase in swim power, and a 4% enhancement in competition performance as a result of a 3-week taper in six male collegiate swimmers (figure 3.7). In keeping with these results, Papoti and colleagues (2007) reported a 3.8% gain in swim force and a 1.6% gain in 200 m time trial performance after an 11-day taper consisting of a 48% reduction in weekly training volume (figure 3.8). Swim force significantly correlated with performance both before and after the taper.

Trinity and colleagues (2006) found that elite swimmers' maximal arm power increased 10% and 12% during a taper, and these gains in power correlated with performance gains of 4.4% and 4.7%. Interestingly, these authors observed that maximal mechanical power displayed a *biphasic response* during the taper, such that approximately 50%, 5%, and 45% of the total increase occurred during the first, second, and third weeks of the taper, respectively. In this study, the biphasic response was reported to be the most common response to the taper among individual swimmers, although there were also "early responders" and "late responders" (figure 3.9). There were also three nonresponders and two swimmers whose power showed a constant increase during the taper.

Prins and colleagues (1991) and Hooper and colleagues (1999) reported unchanged muscular force as a result of a taper, concluding that pretaper force levels were not compromised by the training load undertaken by swimmers. Differences with studies reporting gains in force after a taper may relate to variations in the caliber of the swimmers and the training and tapering programs undertaken.

In cross country and middle-distance runners, Shepley and colleagues (1992) observed an increase in maximal voluntary isometric strength of the knee extensors after both a high-intensity, low-volume taper and a low-intensity, moderate-volume taper, despite

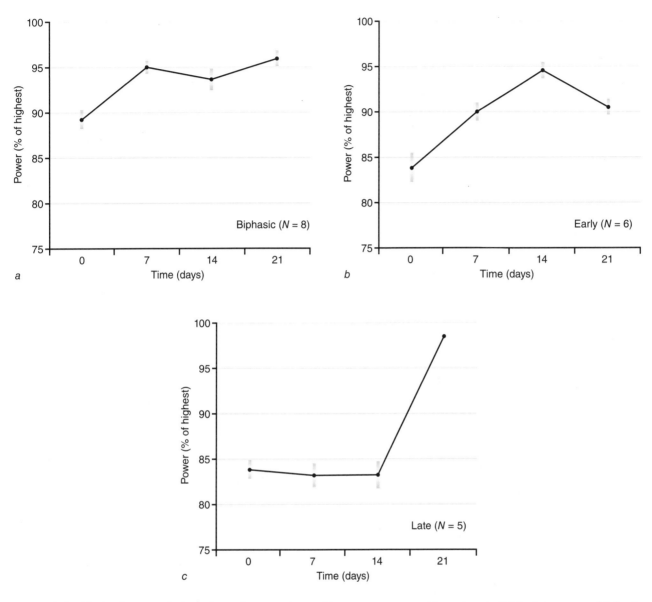

Figure 3.9 Maximal power during a 3-week taper when subjects were grouped based on individual responses: biphasic responders (*n* = 8), early responders (*n* = 6), and late responders (*n* = 5).

Reprinted, by permission, from J.D. Trinity, M.D. Pahnke, E.C. Resse, and E.F. Coyle, 2006, "Maximal mechanical power during a taper in elite swimmers," *Medicine & Science in Sports & Exercise* 38: 1643-1649.

unchanged percentage motor unit activation (figure 3.10). Following percutaneous nerve stimulation, evoked contractile properties of the right knee extensors also improved during both tapers, with peak twitch torque gains amounting to 13% and 19%, respectively (Shepley et al. 1992). Similar positive changes were observed when the athletes performed no training at all for a week, a strategy that resulted in a 3% performance decline. This and other similar findings, such as a 12% gain in peak tethered swimming force that was not accompanied by statistically significant improvements in competition performance during a taper in male and female swimmers (Hooper et al. 1998), or a 2.8% treadmill running performance gain despite a lack of change in leg peak isometric or concentric force (Houmard et al. 1994), suggest that the involvement of muscular force is influenced by many physiological and environmental factors.

In contrast with these findings, the evoked contractile properties (peak twitch torque, time to peak twitch, half relaxation time, and maximum rate of torque development) were unchanged after 4 and 10 days of taper in strength-trained athletes (Gibala et al. 1994). The taper, however, induced an increase in the isometric peak torque of the elbow flexors (7.5% at day 6 and 6.8% at day 10 of the taper), despite unchanged motor unit activation, and increased the concentric peak torque at low velocity (7.7% at day 4 and 2.8% at day 10). These results were attributed to enhanced contractile performance, an increase in neural activation, or both (Gibala et al. 1994). Collegiate cyclists showed improvements in their isokinetic quadriceps strength at 30° and 120° per second after a 2-week step taper, but strength gains were not correlated with improvements in laboratory cycling performance (Martin et al. 1994). Izquierdo and colleagues (2007), studying a group of strength-trained athletes during a 4-week taper, reported 2% gains in 1RM bench press and 3% in maximum concentric 1RM thigh parallel squat, but these gains did not result in functional improvements in the height of a counter-movement jump.

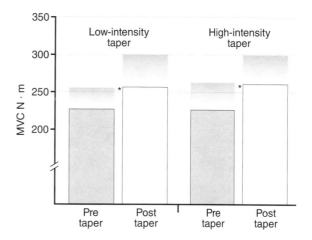

Figure 3.10 Maximal voluntary isometric strength (MVC) of knee extensors of one leg before and after each taper procedure ($N = 9$). Values are mean ± SD. *Significant pre- to posttaper differences, $p < .05$.

Adapted, by permission, from B. Shepley, J.D. MacDougall, N. Cipriano, et al., 1992, "Physiological effects of tapering in highly trained athletes. *Journal of Applied Physiology* 72(Feb): 708. Permission conveyed through Copyright Clearance Center, Inc.

Implications of the Strength and Power Studies

AT A GLANCE

The studies described here suggest that muscular strength and power production are usually suppressed by chronic intensive training but most likely recover during the taper, when the training load is markedly reduced. The mechanisms responsible for the taper-induced improvements in muscular strength and power may be related to local changes in enzymatic activities and single muscle fiber characteristics positively affecting neuromuscular, biomechanical, and metabolic efficiency.

Enzymatic Activities

Neary and colleagues (1992) failed to observe changes in *mitochondrial enzymes* carnitine palmityl transferase, citrate synthase, β-hydroxyacyl CoA dehydrogenase, and cytochrome oxidase, all of which are important for muscle aerobic energy provision, and the key cytoplasmic glycolytic enzyme lactate dehydrogenase after 4 or 8 days of taper. The lack of change could have been related to the small group sizes or a mixture of males and females within each group (Neary et al. 1992). Shepley and colleagues (1992) observed an 18% increase in citrate synthase activity after a 7-day high-intensity, low-volume taper in a group of nine male runners. In light of this finding, and given that citrate synthase is a key enzyme in the process of oxidative energy supply, the investigators attributed part of the laboratory performance gain to an increased capacity to maintain a high rate of oxidative energy production, despite the potentially inhibitory effects of increasing intracellular temperature, hydrogen ion and lactate

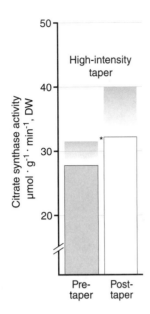

Figure 3.11 Citrate synthase activity for vastus lateralis before and after a high-inensity taper (*N* = 8). Values are mean ± SD and are expressed per unit dry weight. *Significant pre- to posttaper differences, *p* < .05.

Adapted, by permission, from B. Shepley, J.D. MacDougall, N. Cipriano, et al., 1992, "Physiological effects of tapering in highly trained athletes," *Journal of Applied Physiology* 72(Feb): 709. Permission conveyed through Copyright Clearance Center, Inc.

concentration, and superoxide free radicals that occur during high-intensity exercise (Shepley et al. 1992) (figure 3.11).

Muscle Fibers

The taper appears to affect muscle fiber in various ways. Studies on taper-related adaptations in fiber size, metabolic properties, and contractile properties are considered next.

Fiber Size

In male collegiate swimmers, no changes in *Type I muscle fiber* diameter and cross-sectional area were observed after a 3-week taper. On the other hand, Type IIa muscle fiber diameter increased by 11% and cross-sectional area by 24% (Trappe et al. 2001). In male cyclists, a 7-day high-intensity (85-90% of maximal HR), low-volume (progressive reduction from 60 min to 20 min sessions) taper brought about a moderate 6.9% increase in Type I fiber cross-sectional area and a much larger increase, 14%, in *Type II muscle fibers*. When the taper was high volume (60 min sessions) and low intensity (progressive reduction from 85% to 55% of maximal HR), the observed 7.0% and 11% increases in Type I and Type II fibers were not statistically significant (Neary et al. 2003b). In contrast, Harber and colleagues (2004) reported a 4% reduction in Type I fiber diameter in cross country runners during a 4-week taper but unchanged Type IIa diameters.

Metabolic Properties

Only one investigation is available that analyzed the effects of the taper on the metabolic properties of different fiber types. Neary and colleagues (2003b) reported that after a high-intensity, low-volume taper, myofibrillar adenosine triphosphatase (ATPase) increased by 11% and succinate dehydrogenase by 12% in Type I fibers. In Type II fibers, myofibrillar ATPase, succinate dehydrogenase, β-hydroxyacyl CoA dehydrogenase, and cytochrome oxidase increased by 15% to 16%. These changes were accompanied by a 4.3% improvement in a simulated 40 km cycling time trial. After a high-volume, low-intensity taper, cytochrome oxidase (10%) and β-hydroxyacyl CoA dehydrogenase (17%) increased in Type I fibers, but only β-hydroxyacyl CoA dehydrogenase (18%) increased in Type II fibers. The 2.2% improvement in simulated time trial performance was statistically nonsignificant. The succinate dehydrogenase and cytochrome oxidase activities of combined Type I and Type II fibers correlated with simulated 40 km time trial performance after both tapers (figure 3.12). These results illustrate that changes in metabolic properties take place during the taper at the single-fiber level, are more pronounced when a high-intensity taper is performed, and contribute to performance changes observed in whole-body muscle function. Type II fibers seemed to be more responsive to the taper, possibly attributable to their specific contractile properties and a greater potential to increase their oxidative enzyme capacity (Neary et al. 2003b).

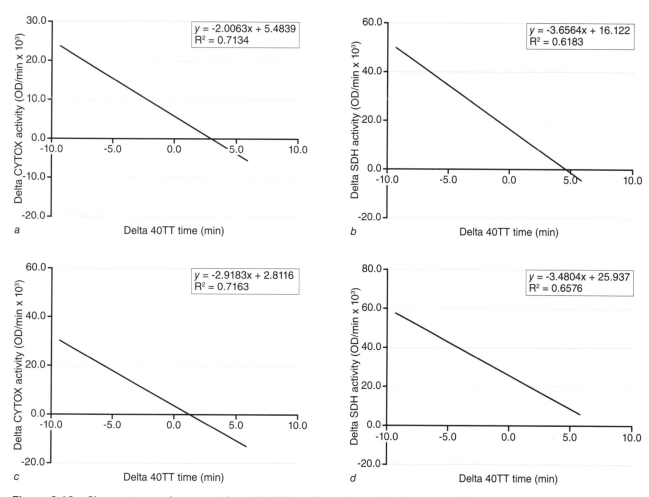

Figure 3.12 Change in cytochrome oxidase (CYTOX) and succinate dehydrogenase (SDH) activity in (a and b) Type I and (c and d) Type II fibers versus 40 km time trial (40TT) time (min) for all groups combined. A negative delta score indicates a faster endurance time but a decreased enzyme activity because values are posttaper minus pretaper.

Adapted, by permission, from J.P. Neary, T.P. Martin, and H.A. Quinney, 2003, "Effects of taper on endurance cycling capacity and single muscle fiber properties," *Medicine & Science in Sports & Exercise* 35: 1879.

Contractile Properties

Neuromuscular adaptations occur at the single-fiber level after tapering. In studying deltoid muscle samples of highly trained collegiate swimmers after a 3-week taper, Trappe and colleagues (2001) observed 30% higher peak isometric force, 67% faster shortening velocity, and 250% higher absolute fiber power in Type IIa muscle fibers. Type I fibers also increased their shortening velocity by 32% (figure 3.13). On average, Type II fibers produced twice as much peak power as Type I fibers before the taper and five times as much peak power after the taper. These observations led the authors to suggest that changes in contractile properties may have been closely related to the observed improvements in whole-muscle strength and power measures after the taper (Trappe et al. 2001).

In cross country runners tapering for 4 weeks, Harber and colleagues (2004) measured a 9% increase in peak force of Type IIa fibers. Conversely, peak force decreased by 9% in Type I fibers. Peak force normalized for fiber size was unaltered for either fiber type by the taper. Single Type I fiber maximal shortening velocity decreased by 17% during the

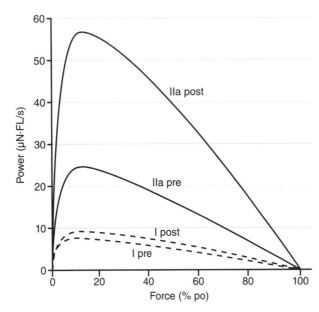

Figure 3.13 Composite force–power curves (absolute power) for Type I and Type IIa posterior deltoid muscle fibers before and after the taper. Measurements were made using isotonic load clamps (force–velocity–power relationships) while maintained at 15 °C.

Reprinted, by permission, from S. Trappe, D. Costill, and R. Thomas, 2001, "Effect of swim taper on whole muscle and single fiber contractile properties," *Medicine & Science in Sports & Exercise* 33: 53.

taper, whereas it did not change in Type IIa fibers. Single fiber absolute power also decreased by 25% in Type I fibers but was unchanged in Type IIa. Power corrected for cell size decreased by 15% during the taper in Type I fibers only. The authors concluded that changes in run training volume and intensity alter myocellular physiology but that the 25% reduction in training volume concomitant with a 67% increase in interval training during the taper was an insufficient training load reduction to elicit robust changes in single-fiber power and run performance (Harber et al. 2004).

As concluded by a recent review article on single-fiber studies and exercise training, the taper has relatively little influence on the size and force characteristics of Type I muscle fibers, but Type IIa fibers seem to be more responsive, showing enhanced contraction performance, as a result of equal or increased cross-sectional area, and increased force and power. In addition, adequate adjustments in training volume and intensity seem to be necessary to elicit positive changes in single-fiber contraction velocity (Malisoux et al. 2007).

AT A GLANCE   Tapered Muscles

Muscular strength and power, usually suppressed by intensive training, increase during the taper, contributing to an athlete's enhanced performance capabilities. Increments in oxidative enzymatic activities of the muscle fibers and positive changes in single muscle fiber size, metabolic potential, and contractile properties are closely related to whole-muscle strength and power measures and performance changes observed in whole-body muscle function.

Immunological Responses

Many aspects of the immune system exhibit a range of responses to acute exercise and prolonged training in athletes preparing for competition: increased leukocyte cell counts, particularly neutrophils and lymphocyte subsets (McCarthy and Dale 1988); decreased functional activity of the neutrophil respiratory burst and natural killer cytotoxicity (Mackinnon 2000, Peake 2002); decreased response to mitogen-induced T-lymphocyte proliferation (Tharp and Preuss 1991); decreased concentration of mucosal immune parameters such as secretory immunoglobulin A (Gleeson et al. 1999, Gleeson and Pyne 2000); impaired delayed-type hypersensitivity response (T-cell function) (Bruunsgaard et al. 1997); and unchanged or increased circulating concentration of cytokines such as the interleukin family, tumor necrosis factor-α, and interferon-δ (Malm 2002).

Despite the high level of clinical and research interest in the effects of exercise and training on immunity, only a limited number of studies have directly examined immunological changes in athletes during the taper phase prior to competition. Most of these studies have focused on the acute effects of exercise on cellular and soluble immune responses immediately postexercise and in the first few hours of recovery. Observations of a temporary suppression in the immune response after intensive exercise led to the creation in the early to mid-1990s of the *J-curve* (Nieman 1993) and *open window* (Hoffman-Goetz and Pedersen 1994) models. More recently, attention has focused on possible mechanisms of immunosuppression, with the T helper lymphocyte subsets T_H1 and T_H2 thought to regulate immunological control (Smith 2003). It appears that the prevailing cytokine pattern elicited by exercise and training activates specific T_H-precursor cells leading to up-regulation of either the T_H1 (cell-mediated immunity) or T_H2 (humoral immunity and antibody production) response (Smith 2003). Future studies are required to fully characterize the time-course of changes in these key immunological control mechanisms during training and the taper.

Although exercise- and training-induced perturbations in immune function of healthy athletes are relatively transient, it is thought that failure to restore baseline levels of immune function after continual upward and downward fluctuations may result in chronic immunosuppression and an increased risk of illness in elite athletes after several years of training (Gleeson 2000, Smith and Pyne 1997). Although evidence of immunosuppression in athletes is indicated indirectly by between-subjects comparison of trained subjects with sedentary individuals, a more direct indication is obtained by analysis of within-subject changes for a given athlete during a specified training interval or taper prior to competition. The key issues for the athletes and for coaches, clinicians, and researchers are the magnitude and duration of immunological changes with training and whether any observed changes manifest as relevant clinical consequences and impair training and competitive performance.

Immune Cells

Studies of athletes in training across a range of sports have examined various immune cell counts and functional activities. These studies generally show relative stability in immunological parameters with little evidence of clinical consequences. Although there is substantial evidence from controlled studies that short segments of intensified training lower resting cell counts (Baj et al. 1994, Fry et al. 1992, Pizza et al. 1995, Verde et al. 1992), observational studies of athletes in training have failed to demonstrate the same findings. One study that experimentally manipulated training volume and intensity in nine well-trained cyclists before an 18-day taper showed improvements in cycling efficiency (6%) and simulated 20 km time trial performance (6%), whereas resting immune status (*lymphocyte* subset counts and incidence of respiratory illness) was unchanged throughout a 10-week training program (Dressendorfer et al. 2002a). Stability in cell counts was also observed in 16 triathletes during 4 weeks of intensified training followed by a 2-week taper (Coutts et al. 2007b) and in 12 young male runners during 40 days of heavy and easy endurance training: Transient reductions were observed in lymphocyte subsets (CD4+ and CD4/CD8 ratio) particularly during times of increased training (high-intensity 1,000 m intervals) compared with increased volume (double normal training volume) (Kajiura et al. 1995).

A small number of studies have directly examined immune cell changes during a taper. National- and international-level swimmers tapering for 4 weeks after 8 weeks of intense training had a decreased percentage of neutrophils after the taper, whereas lymphocytes tended to increase (Mujika et al. 1996c). The increase in lymphocytes correlated positively with the reduction in training volume during taper ($r = .86$). In contrast, a 6-day taper (involving an 80% reduction in high-intensity interval training) in nine male middle-distance runners elicited a modest but statistically significant increase in neutrophils (13%) and granulocytes (11%). However, the magnitude of the observed

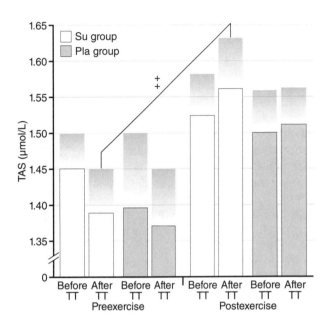

Figure 3.14 Variations in plasma total antioxidant status (TAS) before and after taper training (TT) in pre- and postexercise conditions in supplemented (Su, *n* = 7) and placebo (Pla, *n* = 9) triathletes. ‡$p < .05$ relative to interaction effect between treatment, TT, and exercise.

Adapted, by permission, from I. Margaritis, S. Palazetti, A-S Rousseau, et al., 2003, "Antioxidant supplementation and tapering exercise improve exercise-induced antioxidant response," *Journal of the American College of Nutrition* 22: 152. Permission conveyed through Copyright Clearance Center, Inc.

changes was considered too small to be of immunological significance (Mujika et al. 1996c, Mujika et al. 2002a). A 3-week taper (50% reduction in training volume and intensity after 22 weeks of training at 18-20 hr/week) in 20 collegiate swimmers elicited an increase in total *leukocytes* but a decrease in *B-cell lymphocyte* count (Wilson et al. 1996). However, total *T-cells, neutrophils,* and lymphocyte proliferative responses were unchanged over the same time suggesting that overall immunoprotection was largely unchanged.

It also appears that antioxidant supplementation during the taper enhances plasma antioxidant protection (Margaritis et al. 2003) (figure 3.14), potentially maintaining the delicate balance between oxidant and antioxidant properties of immune cells (Niess et al. 1999). Sixteen male triathletes volunteered for a controlled-training double-blind antioxidant supplementation and taper program. Two weeks of tapered training induced decreases in resting blood *glutathione* concentration, *erythrocyte superoxide dismutase* activity, and plasma antioxidant status but had no effect on *lipoperoxidation* or markers of muscle damage. In seven male distance runners, a 7-day taper did not enhance serum *free radical scavenging* capacity prior to or during exercise (Child et al. 2000). More recently, Vollaard and colleagues (2006) observed no taper-induced changes in resting levels or exercise-induced markers of oxidative stress. Collectively these studies show small transient changes in the distribution of immune cells during the taper that are unlikely to have any substantial clinical consequences.

Immunoglobulins

There has been extensive examination of training-induced changes in soluble immunoproteins in athletes, particularly secretory *immunoglobulin* A (SIgA), which plays a major role in effective specific immunity (Gleeson 2000), and the immunoregulatory cytokines, which are a diverse family of intracellular signaling molecules released by immune cells that exert important influences on inflammatory and immune responses (Suzuki et al. 2002). Some training studies showed that marked reductions in SIgA after acute exercise (Mackinnon and Hooper 1994) and in the latter stages of a prolonged season of training (Gleeson et al. 1995, Gleeson et al. 1999) are associated with an increased risk of upper-respiratory illness. Elite swimmers showed a 4% reduction in SIgA per month over a 7-month season, and values less than 40 mg/L were associated with an increased risk of illness (Gleeson et al. 1999). In contrast, other studies of swimmers and rowers have failed to show this association (Gleeson et al. 2000, Mackinnon and Hooper 1994, Nehlsen-Cannarella et al. 2000, Tharp and Barnes 1990). The interpretation of these disparate findings requires consideration of the inherent biological variability in immune parameters and methodological differences in experimental design, sample collection, and assay techniques. Clearly, further studies are required to resolve these conflicting findings concerning the soluble immune protein response to training and the taper prior to competition.

Cytokines

Many studies have examined the acute effects of exercise on circulating cytokine concentration, but little is known about longitudinal changes in highly trained athletes during training and the taper. *Cytokines* have a role in both the acquired and the innate arms of host defense and can be proinflammatory or anti-inflammatory in nature (Malm 2002). Most of the exercise studies have focused on just three cytokines: interleukin-1α (IL-1α), interleukin-6 (IL-6), and tumor necrosis factor-β. The blood concentration of these cytokines is usually either unchanged or increased in response to acute exercise such as the marathon (Ostrowski et al. 1999). There are also conflicting findings on the influence of exercise intensity and type on cytokine response (Smith et al. 2000). The pattern of postexercise cytokine response (increases in IL-6 and IL-10) to eccentrically biased bench press and leg curl exercise in untrained males was less pronounced and occurred at a later time point (72-144 hr postexercise) (Smith et al. 2000) than after strenuous endurance exercise (Ostrowski et al. 1999). Interpretation of the biological significance is also complicated by the notion that the immunological action of cytokines is regulated by the balance between concentration of the active molecules and their inhibitors rather than their circulating concentration per se. Recent studies have suggested that cytokines such as IL-1 and IL-6 form communication links among systemic circulation, energy metabolism, and skeletal muscle adaptation to exercise (Malm 2002). Studies are required to characterize the time course of cytokine changes during training and the taper in order to fully understand physiological mechanisms underpinning these processes in athletes.

Immune Function   AT A GLANCE

The studies discussed here suggest that athletes should be mindful of excessive loads during peak training but can train with confidence during the taper prior to competition in the knowledge they are unlikely to compromise overall immunological protection. Given the variable findings of existing studies, one-off measures of cellular and soluble immune parameters are unlikely to be informative unless immunosuppression is severe. A multifaceted approach involving systematic monitoring of underlying mucosal and cellular immunity (Gleeson and Pyne 2000); review of clinical, training, and lifestyle factors (Konig et al. 2000); and attention to practical strategies (Pyne et al. 2000) may provide a more effective means of managing the health of athletes during the taper. The sporting community, including athletes, coaches, and physicians, can be assured that maintaining good health during the taper is compatible with elite-level training.

Chapter Summary

Biochemical, hormonal, muscular, and immunological changes may take place in an athlete's body over the course of a taper program, and these changes can sometimes correlate with performance outcomes in competition. One of the most widely used biochemical markers of training stress is blood concentration of creatine kinase, which is usually reduced after a taper, indicating recovery from training stress and muscle damage. Other biochemical indexes have been evaluated during tapering programs, but none of them seems to be useful for coaches and athletes.

Testosterone, cortisol, and the testosterone–cortisol ratio have been thoroughly studied in relation to intensive training and tapering, and it appears that they can provide information on the physiological stress, recovery, and performance capacity of an athlete during the taper. However, these hormonal analyses are invasive and expensive, and performance gains also occur without concomitant changes in these parameters. The 24 hr urinary cortisol–cortisone ratio, plasma and urinary catecholamines, growth hormone, and insulin-like growth factor-I also show promise as tools for monitoring training stress and tapering-induced adaptations, but further research is required before general recommendations can be made regarding the use of these biological markers.

Skeletal muscle is extremely responsive to variable levels of functional demands, and it also adapts to the specific characteristics of tapered training. Muscular strength and power, usually suppressed during intensive training, markedly increase during the taper, contributing to an athlete's enhanced performance capabilities. Local increments in oxidative enzymatic activities, which facilitate aerobic energy provision, and positive changes in single muscle fiber size and metabolic and contractile properties are tightly related to whole-muscle strength and power measures and performance changes observed in whole-body muscle function.

There is very limited research on the effects of the taper on the athlete's immune function, but small transient changes in immune cells, soluble immunoproteins, and cytokines during the taper are unlikely to have marked immunological or clinical significance in most individuals.

chapter 4

Taper-Associated Psychological Changes

It is unlikely that the physiological changes described in chapters 2 and 3 completely explain the performance benefits associated with a successful taper. Because competition performance is the result of a conscious effort (Noakes 2000), it would be a major oversight to obviate the paramount contribution of psychological and motivational factors to posttaper athletic performance. The optimization of an athlete's physiological status resulting from a well-designed tapering strategy is presumably accompanied by beneficial psychological changes, including mood state, perception of effort, and quality of sleep. Reports describing these changes are summarized here, and the chapter is completed with an incursion into the somewhat unexplored territory of the periodization of psychological skills training and its implications for the tapering and peaking training phases.

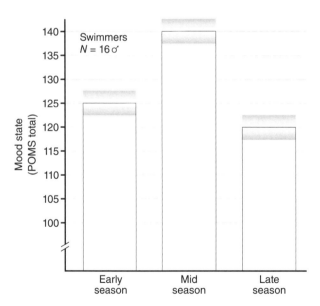

Figure 4.1 Global mood state at the outset, middle, and close of a selected macrocycle.

Reprinted from W.P. Morgan, D.R. Brown, J.S. Raglin, et al., 1987, "Psychological monitoring of overtraining and staleness," *British Journal of Sports Medicine* 21: 109. Reproduced with permission from the BMJ Publishing Group.

Mood State

Considerable scientific evidence indicates that habitual physical activity is associated with positive mental health. However, exercise programs characterized by high levels of intensity, duration, and frequency result in mood disturbances, that is, alterations in an athlete's psychological state (Mondin et al. 1996). Mood states are sensitive to variations in the training load undertaken by athletes (Morgan et al. 1987), and alterations should be expected as a result of a taper where the training load is markedly reduced. Numerous authors have reported mood state changes associated with a precompetition taper (figure 4.1). Most of these reports indicate that tapering induces positive changes in the athlete's mood state, contributing to enhanced performance measures (table 4.1).

Table 4.1 Effects of the Taper on Mood State and Perception of Effort

Study (year)	Athletes	Taper duration, days	Mood state	Perception of effort	Performance measure	Performance outcome, %
Morgan et al. (1987)	Swimmers	28	↑	↓	NR	NR
Raglin et al. (1991)	Swimmers	28	↑	NR	NR	NR
Snyder et al. (1993)	Cyclists	14	NR	↓	NR	NR
Berglund and Säfström (1994)	Canoeists	21	↑	NR	NR	NR
Flynn et al. (1994)	Runners Swimmers	21	↔ ↑	↔ ↓	Treadmill time to exhaustion 23 m, 366 m time trial	↔ ≈3 ↑
Houmard et al. (1994)	Runners	7	NR	↔	5K treadmill time trial	2.8 ↑
Raglin et al. (1996)	Swimmers	28-35	↑	NR	Competition	2.0 ↑
Berger et al. (1997)	Swimmers	7	↑	NR	NR	NR
Taylor et al. (1997)	Swimmers	NR	Deteriorated	NR	Competition	1.3 ↑
Hooper et al. (1998)	Swimmers	14	↑	NR	100 m, 400 m time trial	↔
Berger et al. (1999)	Cyclists	14	↑	NR	4 km simulated pursuit	2.0 ↑
Hooper et al. (1999)	Swimmers	14	↔	NR	100 m time trial	↔
Martin and Andersen (2000)	Cyclists	7	NR	↓	Incremental maximal test	≈6 ↑
Martin et al. (2000)	Cyclists	7	↔	NR	Incremental maximal test	≈6 ↑
Steinacker et al. (2000)	Rowers	7	↑	NR	2,000 m time trial—competition	6.3 ↑
Eliakim et al. (2002)	Handball players	14	↑	NR	4 × 20 m sprint, vertical jump	2.1-3.2 ↑
Margaritis et al. (2003)	Triathletes	14	↑	NR	30 km outdoor duathlon	1.6-3.6 ↑
Neary et al. (2003a)	Cyclists	7	NR	↔	20 km simulated time trial	5.4 ↑
Coutts et al. (2007b)	Triathletes	14	↑	NR	3 km run time trial	3.9 ↑

NR = not reported; ↓ indicates decreased; ↑ indicates increased; ↔ indicates unchanged.

Adapted, by permission, from I. Mujika, S. Padilla, D. Pyne, et al., 2004, "Physiological changes associated with the pre-event taper in athletes," *Sports Medicine* 34: 917.

Studies Using the POMS

Morgan and colleagues (Morgan et al. 1987, Raglin et al. 1991) first described decreased global mood scores computed from the *Profile of Mood States (POMS)* questionnaire in college athletes tapering for 4 weeks. The decrease in global mood scores was associated with decreased levels of perceived fatigue, depression, anger, and confusion. These changes were also accompanied by increased levels of vigor (figure 4.2).

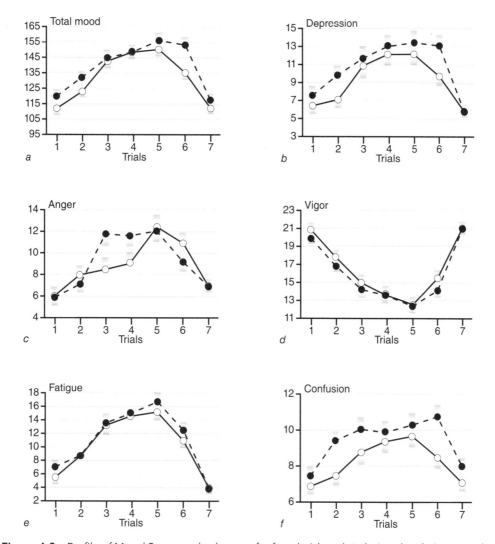

Figure 4.2 Profile of Mood States scale changes for female (closed circles) and male (open circles) swimmers at different times during training.

Reprinted, by permission, from J.S. Raglin, W.P. Morgan, and P.J. O'Connor, 1991, "Changes in mood states during training in female and male college swimmers," *International Journal of Sports Medicine* 12: 586.

Raglin and colleagues (1996) also reported that decreases in mood disturbance were related to reductions in the training load (figure 4.3), with identical effects in both males and females. However, some individual athletes did not respond to the taper, and no declines in tension scores were observed, with values being higher in female than male athletes (Morgan et al. 1987, Raglin et al. 1991). In fact, tension was the only mood variable that remained elevated above baseline following the taper (figure 4.4). It has been speculated that elevated tension probably reflects anxiety provoked by the anticipation of the pending major championship (O'Connor et al. 1989).

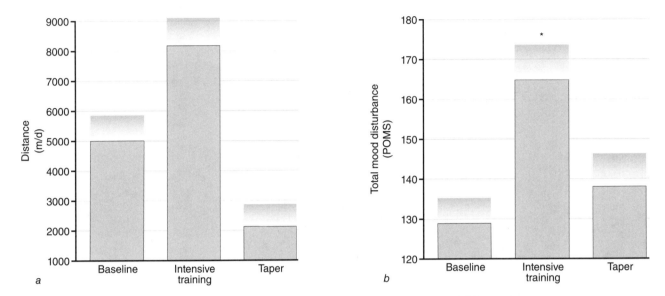

Figure 4.3 Mean (± SE) results across training phases for *(a)* swimming distance and *(b)* mood state.

Reprinted, by permission, from J.S. Raglin, D.M. Koceja, and J.M. Stager, 1996, "Mood, neuromuscular function, and performance during training in female swimmers," *Medicine & Science in Sports & Exercise* 28: 374.

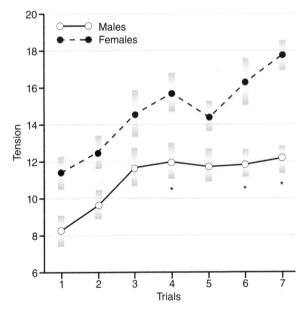

Figure 4.4 Profile of Mood States tension during training for female (closed circles) and male swimmers (open circles).

Reprinted, by permission, from J.S. Raglin, W.P. Morgan, and P.J. O'Connor, 1991, "Changes in mood states during training in female and male college swimmers," *International Journal of Sports Medicine* 12: 587.

In contrast with the previously mentioned findings, Taylor and colleagues (1997) reported gender differences in tapering-induced mood state alterations. Relatively small (1.3%) competition performance gains attained by female swimmers during taper were presumably related to a deterioration in mood state indicated by increases in tension–anxiety (56%), depression–dejection (218%), and confusion–bewilderment (86%) and a 20% decrease in vigor–activity ratings (Taylor et al. 1997).

Flynn and colleagues (1994) reported a 17% reduction in the global mood state of a group of male swimmers after a 3-week taper. A similar 16% decline in total mood disturbance was observed by Raglin and colleagues (1996) in 12 collegiate female swimmers tapering for 4 to 5 weeks. This decline correlated moderately with mean swimming power ($r = .34$), which increased by 20% with the taper. Swimming velocity in competition also improved by 2.0%. Hooper and colleagues (1998) observed reduced tension, depression, and anger after 1 week of taper in state-level swimmers and a 10% lower total mood disturbance after 2 weeks, which resulted in marginal time trial performance gains of 0.2% in 100 m and 0.7% in 400 m events. However, in a subsequent investigation on international-caliber swimmers, these authors did not detect any change in the total mood disturbance after a 2-week taper (Hooper et al. 1999). In another study, young competitive swimmers showed an acute decrease in total mood disturbance after practices that were shorter in duration than usual during a precompetition week of taper. These competitive swimmers reported short-term mood benefits including decreases in scores of depression, confusion, and tension. However, these acute mood benefits during training prior to competition did not appear to be related to subsequent performance in competition (Berger et al. 1997).

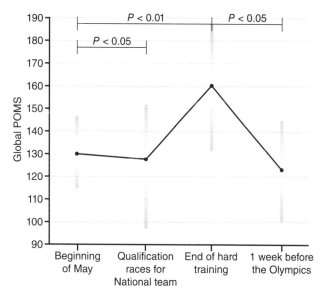

Figure 4.5 Mean (± SD) global profile of mood states scores of world-class canoeists at selected training phases.

Reprinted, by permission, from B. Berglund and H. Säfström, 1994, "Psychological monitoring and modulation of training load of world-class canoeists," *Medicine & Science in Sports & Exercise* 26: 1038.

Total mood disturbance has also been shown to decline by 21% in track cyclists tapering for 2 weeks after a segment of overreaching. At the same time, simulated 4 km pursuit performance improved by 2.0% and mean power output by 2.3%, but no substantial correlations were found between changes in psychological variables and performance changes (Berger et al. 1999). During tapering, the total mood score of the POMS also decreased in nine male and five female world-class canoeists tapering for 3 weeks, so that 1 week before the Olympics this score was of the same magnitude as the basal off-season total mood score of the POMS (Berglund and Säfström 1994) (figure 4.5). Twenty long-distance triathletes' POMS score also decreased by 10% to 12% after 14 days of tapering, during which total training load was progressively reduced by 32% to 46% (Margaritis et al. 2003). In contrast, Martin and colleagues (2000) failed to observe any change in the total mood scores or the specific mood state scores (i.e., tension, depression, anger, vigor, fatigue, confusion) in a group of cyclists tapering for 1 week after 6 weeks of high-intensity interval cycling, despite major differences in the training load and performance assessments at the end of each of these training phases. These authors also indicated that some athletes with relatively large mood disturbances responded well to the taper, whereas others responded poorly, suggesting a low predictability of the POMS for responses to the taper (Martin et al. 2000).

Studies Using Recovery–Stress Measures

A 1-week step taper was enough for 10 world-class male junior rowers showing clear signs of overreaching to recover after 18 days of intense training. Somatic complaints and somatic relaxation assessed with the *Recovery–Stress Questionnaire for Athletes* returned to baseline values during the taper (figure 4.6). Maximal power during an incremental rowing test increased by 2.7%, and a 2,000 m time-trial performance improved by 6.3% (Steinacker et al. 2000). Similarly, Eliakim and colleagues (2002) reported a return to baseline in the self-assessment physical conditioning score during 2 weeks of less intense training in a group of junior handball players, the values of which were reduced during the preceding intensive training.

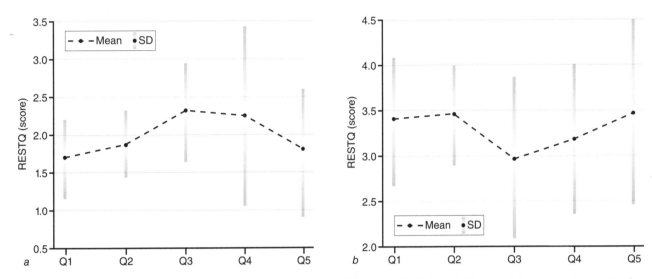

Figure 4.6   Results of the Recovery–Stress Questionnaire for Athletes (RESTQ–Sport). Means of the respective scale after training phases 0 to 5, average and standard deviations. *(a)* RESTQ–Sport Somatic Complaints subscale and *(b)* RESTQ–Sport Somatic Relaxation subscale.

Reprinted, by permission, from J.M. Steinacker, W. Lormes, M. Kellmann, et al., 2000, "Training of junior rowers before world championships. Effects on performance, mood state and selected hormonal and metabolic responses," *Journal of Sports Medicine and Physical Fitness* 40(4): 332.

In a recent investigation on triathletes, Coutts and colleagues (2007b) assessed the athletes' recovery–stress state during 4 weeks of either normal training or intensified training, followed by a 2-week progressive taper. The intensified training group experienced a marked decrease in total stress and an increase in total recovery compared with the normal training group following the taper. The recovery–stress state (total recovery minus total stress) was accordingly affected by changes in the training load. In addition, the stress subscales of Lack of Energy, Physical Complaints, and Fitness/ Injury decreased, whereas the recovery subscale Fitness/Being in Shape increased in the intensified training group.

AT A GLANCE   **How Does the Taper Affect an Athlete's Mood State?**

Coutts and colleagues (2007b) reported that an improved psychological state is an important contributor to posttaper performance gains and that a 2-week taper was sufficient to restore psychological recovery measures to baseline levels.

- The reduced training load associated with the taper facilitates a recovery of an athlete's mood state, which is usually altered during times of intensive training.
- The taper usually induces reductions in perceived fatigue, depression, anger, and confusion.
- The taper enhances the feeling of vigor.
- Tension may not decrease during the taper, attributable to the impending competition.
- All of these psychological markers of reduced stress and enhanced recovery contribute to tapering-induced performance gains.

Perception of Effort

A fatigued and overstressed athlete perceives a given training load as a heavy burden he struggles to deal with. In contrast, a fresh, well-recovered athlete may perceive a similar training load as light and easy. How an athlete perceives a given training bout is his *perception of effort*. The perception of effort during exercise is influenced by a number of physiological and psychological variables (Borg et al. 1987, Noble and Robertson 2000, Watt and Grove 1993), some of which are presumably affected by a taper. The most widely used measure of effort perception is *Borg's Rating of Perceived Exertion (RPE)* (Borg 1970, 1982), which has been shown to change as a result of tapered training (table 4.1).

Studies Using Perceived Exertion Scales

The perception of effort was decreased in swimmers of both genders after a 4-week taper in collegiate athletes (Morgan et al. 1987) (figure 4.7). Flynn and colleagues (1994) reported that participants' RPE while they swam at 90% of preseason $\dot{V}O_2$max decreased from an average value of 14 (somewhat hard to hard) after 2 weeks of hard training to 9 (very light) at the end of the taper. On the other hand, these same authors failed to observe any taper-associated changes in the RPE of eight male cross country runners during treadmill running at 75% of preseason $\dot{V}O_2$max, despite major changes in training loads. The same was true in a study on distance runners performing on a treadmill at 80% $\dot{V}O_2$peak before and after a 7-day taper (Houmard et al. 1994) and on male cyclists performing a simulated 20 km time trial before and after a taper of the same duration (Neary et al. 2003a).

Studies Using Other Measures

The heart rate (HR):RPE relationship could be a more valid marker for monitoring an athlete's response to the taper. Neary and colleagues (2003a) observed a 4.5% decline in the HR:RPE ratio after a 7-day stepwise taper during which training volume was reduced by 50% and performance enhanced by 5.1%. Martin and Andersen (2000) reported a 3.2% decline in the HR:RPE ratio after a 1-week taper in collegiate cyclists, associated with a 6% improvement in a graded exercise to exhaustion (figure 4.8).

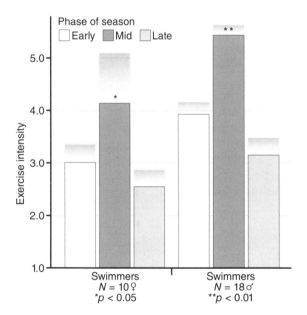

Figure 4.7 Rating of workout intensity during a selected macrocycle.

Reprinted from W.P. Morgan, D.R. Brown, J.S. Raglin, et al., 1987, "Psychological monitoring of overtraining and staleness," *British Journal of Sports Medicine* 21: 109. Reproduced with permission from the BMJ Publishing Group.

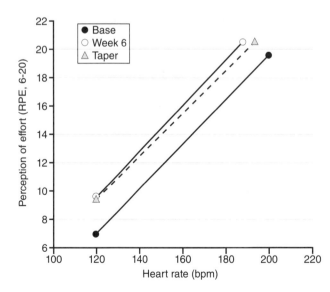

Figure 4.8 The heart rate–perceived exertion relationship for graded exercise tests at baseline, week 6, and taper.

Reprinted, by permission, from D.T. Martin and M.B. Andersen, 2000, "Heart rate-perceived exertion relationship during training and taper," *Journal of Sports Medicine and Physical Fitness* 40(3): 205.

Change in the HR:RPE relationship during 6 weeks of high-intensity interval training was a powerful predictor of performance responses to the taper. Subjects who demonstrated the greatest decrease in HR for a given RPE tended to have the greatest performance increases in response to the taper ($r = .72$), confirming the usefulness of the HR:RPE relationship (Martin and Andersen 2000).

Using a similar approach, Snyder and colleagues (1993) proposed that the submaximal blood lactate concentration (HLa):RPE ratio is a valid physiological–psychological index of fatigue. During 2 weeks of recovery training, the values for the HLa:RPE ratio, which were reduced as a result of 2 previous weeks of high-intensity interval training, returned to normal baseline levels.

AT A GLANCE | **How Does the Taper Affect an Athlete's Perception of Effort?**

The rating of perceived exertion (RPE) can be a useful tool to gauge how an athlete is responding to the taper. The following are ways in which the athlete's RPE is usually correlated with the psychological and physiological effects of the taper.

- The reduced training load associated with the taper facilitates a recovery of an athlete's RPE, which is usually altered during preceding times of intensive training.
- The HR:RPE relationship could be a valid marker of an athlete's response to the taper.
- The relation between submaximal blood lactate concentration and RPE could also be a valid physiological–psychological index of fatigue and recovery.
- A reduced perception of effort is usually associated with performance gains during the taper.

Quality of Sleep

Sleep is a compensatory mechanism following catabolic processes of daytime activity, and sleep disturbance is often associated with excessive training loads and overtraining, defined as "an imbalance between training and recovery, exercise and exercise capacity, stress and stress tolerance" (Lehmann et al. 1993, p. 854). Given that tapering strategies are characterized by reduced training loads, it seems plausible that sleep quality could also be affected by the taper.

Forcing habitual exercisers to spend a sedentary day modifies sleep patterns and body temperature. With reduced exercise load, slow-wave sleep pressure is reduced, resulting in lower levels of slow-wave sleep and increased rapid eye movement sleep (Hague et al. 2003). The most in-depth investigation dealing with sleep patterns during tapering in athletes indicates that sleep-onset latency, time awake after sleep onset, total sleep time, and rapid eye movement sleep time were unchanged during the taper in female swimmers. On the other hand, slow-wave sleep, which represented 31% of total sleep time during peak training, was reduced to 16% following the taper, suggesting that the need for restorative slow-wave sleep is reduced with reduced physical demand. However, the number of movements during sleep was reduced by 37% after the taper, indicating less sleep disruption in comparison with previous times of higher training loads (Taylor et al. 1997) (table 4.2). Hooper and colleagues (1999) reported a slightly improved quality of sleep in seven highly competitive female swimmers after 2 weeks of tapering for the Australian national championships. The measurements were taken at three different phases of their swimming training program: onset of training, peak, and taper. The interval between each set of measurements was approximately 3 months.

Table 4.2 Mean ± Standard Deviation for Selected Sleep Measures From Seven Athletes

	Onset	Peak	Taper	Onset–Peak	Peak–Taper	Onset–Taper
				SIGNIFICANCE LEVEL		
Total recording time, min	470.3 ± 32.0	488.7 ± 57.3	471.1 ± 15.2	NS	NS	NS
Total sleep time, min	451.1 ± 29.7	472.1 ± 60.8	458.4 ± 14.1	NS	NS	NS
Sleep-onset latency, min	19.3 ± 3.5	16.6 ± 5.6	16.9 ± 8.8	NS	NS	NS
REM-onset latency, min	73.3 ± 37.6	64.3 + 16.0	60.7 ± 8.9	NS	NS	NS
Wake after sleep onset, min	4.4 ± 6.4	9.0 ± 13.8	9.1 ± 5.3	NS	NS	NS
Stages 1 and 2, % TST	49.2 ± 7.9	44.5 ± 6.4	59.5 ± 3.6	NS	$p < .05$	NS
Slow-wave sleep, % TST	28.7 ± 5.1	29.4 ± 3.2	18.4 ± 4.3	NS	$p < .05$	$p < .01$
REM sleep, % TST	16.7 ± 4.0	18.3 ± 4.0	17.2 ± 2.5	NS	NS	NS
Shift to awake	5 ± 7	4 ± 3	3 ± 3	NS	NS	NS
Total no. of movements	46 ± 26	49 ± 30	31 ± 19	NS	$p < .01$	$p < .05$

NS = not significant; REM = rapid eye movement; TST = total sleep time.

Reprinted, by permission, from S.R. Taylor, G.G. Rogers, and H.S. Driver, 1997, "Effects of training volume on sleep, psychological, and selected physiological profiles of elite female swimmers," *Medicine & Science in Sports & Exercise* 29: 688-693.

Tapering and Sleep

AT A GLANCE

Sleep is the most important recovery mechanism for an athlete, but its quality is often altered during intensive training. During tapering, on the other hand, athletes tend to sleep better, which along with the reduced physical demand associated with the taper facilitates recovery and optimizes competition performance.

Periodized Psychological Skills Training

The classic concept of periodized training, by which the annual plan is divided and structured into different training phases to lead to optimal performance for the main competition, has recently been embraced by sport psychology. Professor Gloria Balague of the University of Illinois proposed a model for training psychological skills following the notion of periodized training (Balague 2000). Professor Balague states that psychological skills relevant to sport performance can be learned and improved by training in the same manner as can physical skills. In her model, she first defines the goals of psychological skills training, which are to help athletes reach their potential by maximizing learning and performance and, more specifically, to increase consistency by increasing the athletes' control over their performance. She moves on to the classification of psychological skills in three major groups: foundation skills (including motivation, self-awareness, and interpersonal skills), performance skills (including self-efficacy, self-regulation of physical arousal, cognitive and emotional self-control, and self-regulation of attention), and facilitative skills (including lifestyle management and media skills).

The next step for successful periodization of psychological skills involves assessment of the requirements of the sport or event and evaluation of the athlete in terms of psychological skills attributes, deficits, strengths, and weaknesses through interviews, observation, and the use of psychometric testing. Finally, the specific notion of periodized psychological skills training can be introduced, which must follow closely the demands of each specific training phase and the goals an athlete is targeting during each of these phases.

Using the example of psychological skills training for long and triple jump, the author describes the psychological requirements and interventions to match the technical, tactical, and performance goals of each training phase (table 4.3). In this example, the main goal of the precompetitive phase is to finalize the jump to gain consistency. The psychological goal would be to help the athlete to identify the elements associated with good performance so that they can be reproduced. These include optimal level of arousal, optimal focus of attention, and cognitive and emotional control. The main interventions during this phase target learning to focus attention, developing precompetitive routines, and identifying the most adaptive behaviors between jumps. Some other tasks to accomplish during this phase include identifying self-dialogue, effective cue words, and images that evoke energy, rhythm, and confidence (Balague 2000).

During the main competitive phase, the athlete should be prepared to perform independently and consistently and be able to adjust to changing conditions. From a psychological standpoint, the athlete will have to trust his or her skills and be flexible

Table 4.3 Preparation Plan for Horizontal Jumps

AUG	SEP	OCT	NOV	DEC	JAN	FEB	MAR	APR
General Preparation ⟶			Specific Prep			Pre-Comp		Main Comp Transit.
Conditioning phase		**Technical work**		**Tactical work**		**Performance**		**Eval/rest**
Strength, speed, endurance		Jump phases emphasized		Consistency		Self-management		Evaluation
Rhythm		Approach run		Coordination of jump The board/fouls				
Psychological Requirements								
Motivation, pain/ fatigue toler.		Kinesthetic control		Optimal arousal		Trust		Evaluation
Self-efficacy		Awareness of jump phases		Optimal focus		Flexibility		Restoration
Self-awareness		Increased efficacy		Cognitive/emot. control		Emotion control		Self-efficacy
Interventions								
Goal setting Relaxation training Imagery training		Visualization Rhythm work Feedback on improvement		Attention focus work Compet. routines Cognitive restruct.		Comp. plan Comp. eval. Anticipation		Evaluation Perspective Letting go
Cognitive awareness		Kinesth. awareness exercises		Board/confidence exer.		Planning		Self-care
Attentional style eval.				Relaxing/energizing cues				

Reprinted, by permission, from G. Balague, 2000, "Periodization of psychological skills training," *Journal of Science and Medicine in Sport* 3(3): 234.

enough to anticipate and react in the event of unexpected circumstances. The interventions should target the development of a solid competition plan and the postcompetition evaluation of psychological focus and performance, so the athlete can make minor adjustments if necessary. Media skills training could also be required during this phase (Balague 2000).

Benefits of Psychological Skills Training   AT A GLANCE

Given that competition performance is the result of a conscious effort, positive psychological and motivational changes taking place during the taper, including an enhanced mood state, a reduced perception of effort, and an improved quality of sleep, can make a worthwhile contribution to athletic performance. Coaches and athletes also should be aware of the potential benefits to be derived from psychological training techniques and make the best possible use of these to optimize athletes' psychological status in the lead-up to major competitions.

Chapter Summary

Psychological and motivational factors play a key role in attaining optimal competition performance, and a taper can help fine-tune the athlete's psychological state in the lead-up to a major event. During times of intensive training, an athlete's mood state is often negatively affected by feelings of depression, anger, fatigue, and confusion and low levels of vigor. Fortunately, research shows that this situation can be reverted by reducing the training load during the taper.

The taper is often associated with reduced levels of somatic complaints and stress but increased relaxation and recovery in previously overreached athletes, and these changes are generally paralleled by performance improvements. As a consequence, a training load that felt difficult and exhausting before the taper is perceived as lighter and easier at the end of it. A better quality of sleep during the taper also contributes to a better balance between training and recovery, facilitating both physical and psychological preparedness for competition.

The physical and psychological preparedness of an athlete can be systematically monitored with simple tools such as mood state, perception of effort, recovery–stress, and sleep quality questionnaires.

part II

Tapering and Athletic Performance

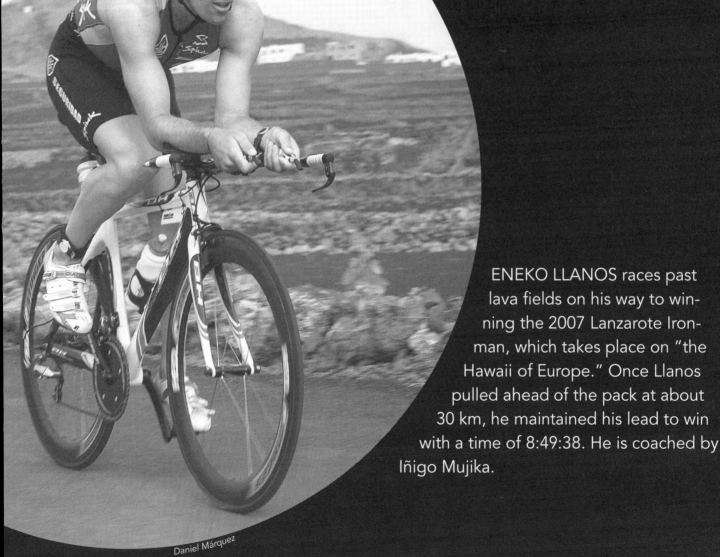

ENEKO LLANOS races past lava fields on his way to winning the 2007 Lanzarote Ironman, which takes place on "the Hawaii of Europe." Once Llanos pulled ahead of the pack at about 30 km, he maintained his lead to win with a time of 8:49:38. He is coached by Iñigo Mujika.

Daniel Márquez

Coaches and sport scientists who prepare elite athletes for major competitions are well aware of the performance-enhancing potential of a well-designed taper. This awareness is based on both anecdotal reports from successful competitions and a wealth of peer-reviewed publications describing the beneficial performance consequences of a taper following intensive training. Moreover, the relationship between the reduced training load during the taper and performance benefits is fairly well established, allowing investigators to make training recommendations to optimize the effects of preevent tapering strategies (Houmard 1991, Houmard and Johns 1994, McNeely and Sandler 2007, Mujika 1998, Mujika and Padilla 2003a, Neufer 1989).

The chapters in part II will enable the readers to take advantage of the available practical knowledge on optimal taper designs. Indeed, this part comprehensively addresses the issues of

- how, by how much, and for how long training should be reduced prior to competition, and what environmental variables may interact with the taper in international competition;
- the size of the performance gains that can be achieved through the taper, the variables that may or may not affect these performance benefits, and what these benefits mean in terms of competition placing;
- the modeled effects of tapering on athletes' fitness and fatigue profiles, and how these affect performance capabilities; and
- the unique aspects to consider when preparing for team sport competition, whether it is peaking for a regular season or for a major tournament.

Part II will thus help readers confidently design successful taper programs for their particular sport.

chapter 5

Reduction
of the Training Load

The training load or training stimulus in competitive sport can be described as a combination of training intensity, volume, and frequency (Wenger and Bell 1986). This training load is markedly reduced during the taper in an attempt to reduce accumulated fatigue, but reduced training should not be detrimental to training-induced adaptations. An insufficient training stimulus could bring about a partial or complete loss of training-induced anatomical, physiological, and performance adaptations, in other words, detraining (Mujika and Padilla 2000). Therefore, athletes and their coaches must determine the extent to which the training load can be reduced at the expense of the training variables while retaining or slightly improving adaptations and avoiding a detrained state. This chapter establishes the scientific bases for successfully reducing precompetition training loads, helping individual athletes, coaches, and sports scientists achieve the optimum training mix during the taper, leading to peak performances at the desired point during the season.

Intensity

In the third and last part of a now-classic series of studies, Hickson and colleagues (1985) demonstrated that training intensity is essential for maintaining training-induced adaptations during times of reduced training in moderately trained individuals. These authors trained subjects in cycling and in treadmill running. For 10 weeks the subjects trained 6 days/week for 40 min/day at intensities approaching their $\dot{V}O_2$max. For the next 15 weeks the group was divided so that some trained at two thirds of the original $\dot{V}O_2$max level and others at only one third of the original $\dot{V}O_2$max level but with the frequency and duration of the training maintained at the original levels.

The authors reported that gains attained during the 10 weeks of intensive training in aerobic power as measured by $\dot{V}O_2$max, endurance as measured by cycling time to exhaustion, and cardiac growth were not maintained when intensity was reduced during the 15 weeks. The decreases in $\dot{V}O_2$max were, not surprisingly, greater with a two-thirds reduction in work rate than with a one-third reduction in work rate. After training was reduced, the subjects who reduced their intensity by one third experienced a 21% loss of performance and the subjects who reduced by two thirds had a 30% loss of performance. These results show that training intensity plays a key role in regulating the maintenance of increased aerobic power. Figures 5.1 and 5.2 show both the original training effect as measured by $\dot{V}O_2$max (figure 5.1) or cycling to exhaustion (figure 5.2) and the loss of that effect when training intensity was reduced by either one third or two thirds.

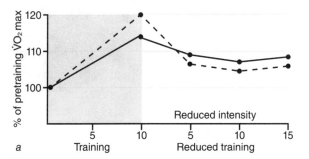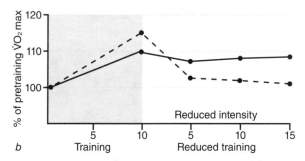

Figure 5.1   Effects on $\dot{V}O_2$max levels of subjects trained for 10 weeks at intensities approaching their $\dot{V}O_2$max followed by 15 weeks of reduced training. Both frequency and duration were maintained whereas training intensity levels were reduced by either one third (solid lines) or two thirds (broken lines). *(a)* Effects on cycling $\dot{V}O_2$max and *(b)* effects on treadmill running $\dot{V}O_2$max.

Adapted, by permission, from R.C. Hickson, C. Foster, M.L. Pollock, et al., 1985, "Reduced training intensities and loss of aerobic power, endurance, and cardiac growth," *Journal of Applied Physiology* 58(Feb): 497. Permission conveyed through Copyright Clearance Center, Inc.

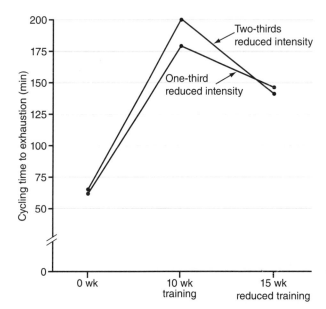

Figure 5.2   Effects on cycling time to exhaustion of 10 weeks of training and 15 weeks of reduced training at one third or two thirds reduction in training intensities. Within each group, all points are significantly different from each other.

Adapted, by permission, from R.C. Hickson, C. Foster, M.L. Pollock, et al., 1985, "Reduced training intensities and loss of aerobic power, endurance, and cardiac growth," *Journal of Applied Physiology* 58(Feb): 495. Permission conveyed through Copyright Clearance Center, Inc.

The paramount importance of training intensity for maintaining training-induced physiological and performance adaptations has also been demonstrated in intervention studies performed with highly trained athletes. In a study that is mentioned several times in this book, Shepley and colleagues (1992) compared some of the physiological and performance effects of a high-intensity, low-volume taper; a low-intensity, moderate-volume taper; and a rest-only taper in middle-distance runners. Total blood volume, red cell volume, citrate synthase activity, muscle glycogen concentration, muscle strength, and running time to fatigue were optimized only with the high-intensity, low-volume taper. In this respect, the major influence of training intensity on the retention or improvement of training-induced adaptations could be explained by its role in the regulation of concentrations and activities of fluid retention hormones (Convertino et al. 1981, Mujika 1998). In addition, Mujika and colleagues (2000) reported that high-intensity interval training during the taper correlated positively with the percentage change in circulating testosterone levels in a group of well-trained middle-distance runners tapering for 6 days, which could facilitate the recovery processes of the athletes in the lead-up to competition.

In a recent meta-analysis of the available tapering literature on highly trained athletes, Bosquet and colleagues (2007) concluded that the training load should not be reduced

Table 5.1 Effect of Decreasing or Not Decreasing Training Intensity During the Taper on Overall Effect Size for Taper-Induced Changes in Performance

Decrease in training intensity	Overall effect size, mean (95% confidence interval)	n	p
Yes	−0.02 (−0.37, 0.33)	63	.91
No	0.33 (0.19, 0.47)	415	.0001

Data reprinted, by permission, from L. Bosquet, J. Montpetit, D. Arvisais, et al., 2007, "Effects of tapering on performance: A meta-analysis," *Medicine & Science in Sports & Exercise* 39: 1358-1365.

Table 5.2 Effect of Decreasing or Not Decreasing Training Intensity During the Taper on Overall Effect Size for Taper-Induced Changes in Swimming, Running, and Cycling Performance

	SWIMMING		RUNNING		CYCLING	
Decrease in training intensity	Mean (95% confidence interval)	n	Mean (95% confidence interval)	n	Mean (95% confidence interval)	n
Yes	0.08 (−0.34, 0.49)	45	−0.72 (−1.63, 0.19)	10	0.25 (−0.73, 1.24)	8
No	0.28 (0.08, 0.47)*	204	0.37 (0.09, 0.66)*	100	0.68 (0.09, 1.27)**	72

*$p \leq .01$; **$p \leq .05$.

Data reprinted, by permission, from L. Bosquet, J. Montpetit, D. Arvisais, et al., 2007, "Effects of tapering on performance: A meta-analysis," *Medicine & Science in Sports & Exercise* 39: 1358-1365.

at the expense of training intensity during a taper, because intensity is considered a key parameter in maintaining training-induced adaptation during this time (Mujika 1998, Mujika et al. 2004). The meta-analysis approach of Bosquet and colleagues allowed them to assess the effect of decreasing or not decreasing training intensity during the taper on the overall effect size of taper-induced changes in performance, both in general (table 5.1) and separately in swimming, running, and cycling (table 5.2). In every case, not decreasing training intensity had a preferable impact on sport performance.

In their reviews, other authors have underlined the importance of training intensity during a taper (Houmard and Johns 1994, Kubukeli et al. 2002, Mujika 1998, Neufer 1989). Recently, McNeely and Sandler (2007) indicated that the race-pace training performed in the final days before competition is as important in psychological terms as physiological terms. Race-pace intervals in the final training sessions give athletes feelings of speed, power, and confidence but should leave an athlete feeling energized rather than fatigued (McNeely and Sandler 2007). These authors also provide an example of the final week of a taper successfully used with rowers prior to Olympic Games and World Championships (table 5.3).

Mujika and colleagues (1996a) reported on the weekly swimming distance performed at different intensity levels by 18 national- and international-caliber swimmers in the 4 weeks leading up to the taper and during each of the three tapers performed by the athletes during a season. As shown in table 5.4, during the first taper of the season, lasting 3 weeks, the swimmers reduced their weekly distance swum at moderate and maximal intensity. During the second taper, lasting 4 weeks, weekly distances at all levels of intensity were reduced. During the third taper, which lasted 6 weeks, only moderate- and high-intensity swimming was reduced.

Table 5.3   Final Taper Week for a Rower Training More Than 15 Hr/Week

Monday	Tuesday	Wednesday	Thursday	Friday	Saturday	Sunday
40 min easy steady state	4 × 5 min at anaerobic threshold, 10 min active rest between	5 × 4 min at anaerobic threshold, 30 min easy steady state	Off	4-6 × 3 min at race pace, 5 min easy steady state between	4 × 2 min at race pace, 5 min between	Race

Reprinted, by permission, from E. McNeely and D. Sandler, 2007, "Tapering for endurance athletes," *Strength and Conditioning Journal* 29(5): 22.

Table 5.4   Swimming Training Before and During Each of the Three Tapers (Mean ± Standard Deviation)

Training intensity	TAPER 1		TAPER 2		TAPER 3	
	Before (4 wk)	During (3 wk)	Before (4 wk)	During (4 wk)	Before (4 wk)	During (6 wk)
I	27.28 ± 6.79	25.55 ± 5.20	25.96 ± 6.68	20.05 ± 3.96*	22.34 ± 5.24	17.76 ± 4.78
II	5.78 ± 2.62	1.12 ± 0.90**	4.95 ± 2.32	0.65 ± 0.71**	2.79 ± 1.41	1.47 ± 0.87
III	2.55 ± 1.10	0.89 ± 0.61**	2.69 ± 0.82	1.01 ± 0.52**	3.10 ± 1.09	1.55 ± 0.51**
IV	1.27 ± 0.32	1.02 ± 0.30	1.32 ± 0.42	0.92 ± 0.26*	1.43 ± 0.39	1.09 ± 0.31*
V	0.43 ± 0.11	0.29 ± 0.07*	0.38 ± 0.08	0.25 ± 0.06*	0.28 ± 0.09	0.27 ± 0.18

Intensity I ≈ 2 mmol/L; intensity II ≈ 4 mmol/L; intensity III ≈ 6 mmol/L; intensity IV ≈ 10 mmol/L; intensity V ≈ sprint training. $*p < .05$; $**p < .001$.

Reprinted, by permission, from I. Mujika, T. Busso, L. Lacoste, et al., 1996, "Modeled responses to training and taper in competitive swimmers," *Medicine & Science in Sports & Exercise* 28: 251-258.

Stewart and Hopkins (2000) provided an elaborate description of training practices of 24 swim coaches and 185 swimmers surveyed over a summer and winter season in New Zealand. As shown in figure 5.3, interval training intensity and rest duration of interval workouts increased during the taper for sprinters (50 and 100 m) and middle-distance (200 and 400 m) specialists, whereas interval distance decreased.

AT A GLANCE   Maintaining Intensity During the Taper

The training load is markedly reduced during a taper so that athletes recover from intensive training and feel energized before major events. With a reduced training load, however, there may be a risk of detraining. Research clearly shows that to avoid this risk, training intensity should be maintained during the taper, because this training variable is key to retaining aerobic power, circulating anabolic hormones, and feelings of speed and power. In addition, high-quality training during the taper can further enhance physiological and performance adaptations. Conversely, if training intensity is diminished, some training-induced adaptations may be lost, leading to suboptimal competition performance.

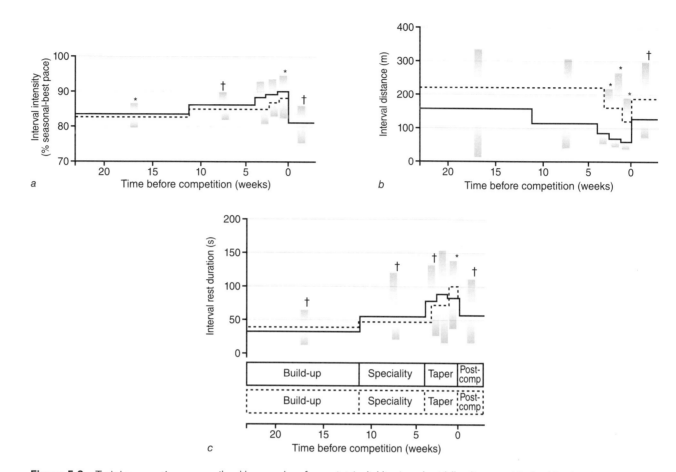

Figure 5.3 Training practices prescribed by coaches for sprint (solid line) and middle-distance (dashed line) swimmers. Data are the means of the summer and winter seasons for 24 coaches. Error bars represent standard deviations. The taper was subdivided into three phases: start, middle, and end. *$p < .05$; † $p < .001$: significantly different from value in previous phase.

Adapted from A.M. Stewart and W.G. Hopkins, 2000, "Seasonal training and performance of competitive swimmers," *Journal of Sports Sciences* 18(11): 878. By permission of the publisher (Taylor & Francis Ltd., http://www.tandf.co.uk/journals).

Volume

Hickson and colleagues (1982) showed that subjects moderately trained in either cycling (figure 5.4*a*) or treadmill running (figure 5.4*b*) appear to retain gains attained through 10 weeks of training during 15 subsequent weeks during which training time was reduced by one or two thirds (from 40 min/day to either 26 or 13 min/day). The gains were achieved and maintained in $\dot{V}O_2$max, peak blood lactate concentrations, calculated left ventricular mass, and short-term endurance (exercise to exhaustion at an intensity corresponding to the maximal oxygen uptake). Long-term endurance (cycling time to exhaustion at approximately 80% of $\dot{V}O_2$max) was also maintained by the group that reduced duration by one third while maintaining intensity (cycling at approximately 80% $\dot{V}O_2$max). However, there was a significant loss of performance by the 13 min group (two-thirds reduction in duration), as average time to exhaustion decreased by 10% (figure 5.5; Hickson et al. 1982).

Standardized training volume reductions of 50% to 70% have been shown to be a valid approach to retain or slightly improve training-induced adaptations in well-trained runners (Houmard et al. 1989, Houmard et al. 1990a, Houmard et al. 1990b, McConell et al. 1993) and cyclists (Martin et al. 1994, Rietjens et al. 2001). Conversely, progressive training reductions of up to 85% have been reported to bring about various significant, performance-enhancing physiological changes. Mujika and colleagues (2000) compared the effects of progressive 50% or 75% training volume reductions during a 6-day taper

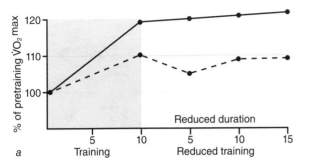
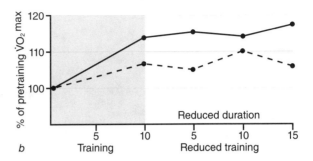

Figure 5.4 Effects of 10 weeks of training and 15 weeks of reduced training at one-third (solid lines) and two-thirds (broken lines) reductions in duration (volume) on (a) cycling and (b) treadmill running maximum oxygen uptake ($\dot{V}O_2$max).

Adapted, by permission, from R.C. Hickson, C. Foster, M.L. Pollock, et al., 1985, "Reduced training intensities and loss of aerobic power, endurance, and cardiac growth," *Journal of Applied Physiology* 58(Feb): 497. Permission conveyed through Copyright Clearance Center, Inc.

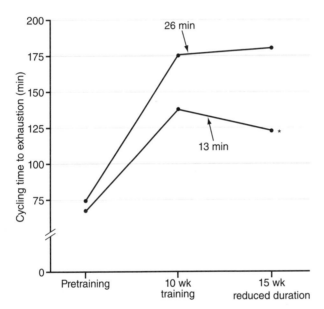

Figure 5.5 Effects of 10 weeks of training and 15 weeks of reduced training at one-third (26 min) and two-thirds (13 min) reductions on cycling time to exhaustion while retaining original intensity. *Significantly different from 10 weeks of training, p < .05.

Reprinted, by permission, from R.C. Hickson, C. Kanakis, Jr., J.R. Davis, et al., 1982, "Reduced training duration effects on aerobic power, endurance, and cardiac growth." *Journal of Applied Physiology* 53(July): 228. Permission conveyed through Copyright Clearance Center, Inc.

in middle-distance runners and concluded that the 75% reduction was a more appropriate strategy to optimize adaptations, because the 75% volume reduction group improved 800 m performance by 0.95%, whereas the 50% reduction group went 0.40% slower after the taper. The investigators also found a negative correlation between the distance of low-intensity continuous training and the percentage change in circulating testosterone during the taper (see figure 3.2).

In a similar group of runners, Shepley and colleagues (1992) found better physiological and performance results with a low-volume taper than with a moderate-volume taper, and in competitive swimmers, a positive relationship has been observed between performance gains and the percentage reduction in training volume during a 3-week taper (figure 5.6, Mujika et al. 1995).

In their 2007 meta-analysis, Bosquet and colleagues analyzed the influence of training volume on taper-induced changes in general sport performance (table 5.5) and in different modes of human locomotion (table 5.6).

In agreement with previous suggestions (Houmard and Johns 1994, Mujika and Padilla 2003a), the study of Bosquet and colleagues (2007) confirmed that performance gains were highly sensitive to reductions in training volume, which were computed as the area under the training volume–time curve. Maximal performance gains were obtained with training volume reductions of 41% to 60% of pretaper training, although performance benefits can be attained with smaller and also bigger training volume decrements. The dose–response curve for the overall effect of percentage decrement in training volume on performance can be seen in figure 5.7.

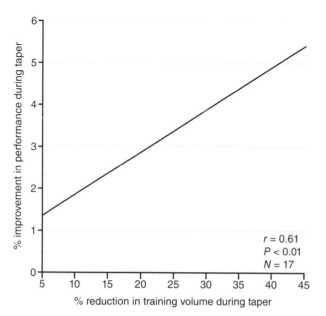

Figure 5.6 Relationship between the improvement in performance during a 3-week taper and the percentage reduction in training volume during taper (mean pretaper weekly volume vs. mean weekly volume of the 3 weeks of taper) in a group of elite swimmers.

Adapted, by permission, from I. Mujika, J.C. Chatard, T. Busso, et al., 1995, "Effects of training on performance in competitive swimming," *Canadian Journal of Applied Physiology* 20: 401.

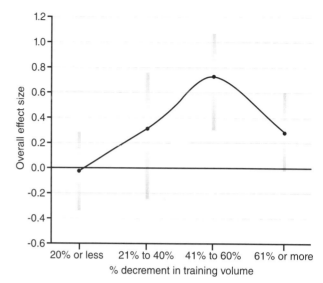

Figure 5.7 Dose–response curve for the effect of percent decrement in training volume during the taper on performance.

Reprinted, by permission, from L. Bosquet, J. Montpetit, D. Arvisais, et al., 2007, "Effects of tapering on performance: A meta-analysis," *Medicine & Science in Sports & Exercise* 39: 1358-1365.

Table 5.5 Effect of Decreasing Training Volume During the Taper on Overall Effect Size for Taper-Induced Changes in Performance

Decrease in training volume	Overall effect size, mean (95% confidence interval)	*n*	*p*
≤20%	−0.02 (−0.32, 0.27)	152	.88
21-40%	0.27 (0.04, 0.49)	90	.02
41-60%	0.72 (0.36, 1.09)	118	.0001
≥60%	0.27 (−0.03, 0.57)	118	.07

Data reprinted, by permission, from L. Bosquet, J. Montpetit, D. Arvisais, et al., 2007, "Effects of tapering on performance: A meta-analysis," *Medicine & Science in Sports & Exercise* 39: 1358-1365.

Table 5.6 Effect of Decreasing Training Volume During the Taper on Overall Effect Size for Taper-Induced Changes in Swimming, Running, and Cycling Performance

	SWIMMING		RUNNING		CYCLING	
Decrease in training volume	Mean (95% confidence interval)	*n*	Mean (95% confidence interval)	*n*	Mean (95% confidence interval)	*n*
≤20%	−0.04 (−0.36, 0.29)	72	No data available		0.03 (−0.62, 0.69)	18
21-40%	0.18 (−0.11, 0.47)	91	0.47 (−0.05, 1.00)**	30	0.84 (−0.05, 1.74)**	11
41-60%	0.81 (0.42, 1.20)*	70	0.23 (−0.52, 0.98)	14	2.14 (−1.33, 5.62)	15
≥60%	0.03 (−0.66, 0.73)	16	0.21 (−0.14, 0.56)	66	0.56 (−0.24, 1.35)	36

*p ≤ .01; **p ≤ .10.

Data reprinted, by permission, from L. Bosquet, J. Montpetit, D. Arvisais, et al., 2007, "Effects of tapering on performance: A meta-analysis," *Medicine & Science in Sports & Exercise* 39: 1358-1365.

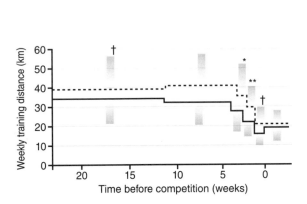

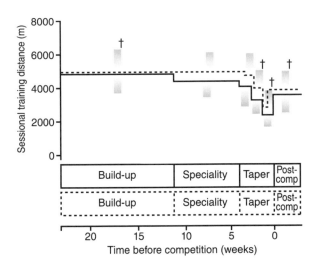

Figure 5.8 Training practices prescribed by coaches for sprint (solid line) and middle-distance (dashed line) swimmers. Data are the means of the summer and winter seasons for 24 coaches. Error bars represent standard deviations. The taper was subdivided into three phases: start, middle, and end. *p < .05, **p < .01, †p < .001: significantly different from value in previous phase.

Adapted from A.M. Stewart and W.G. Hopkins, 2000, "Seasonal training and performance of competitive swimmers," *Journal of Sports Sciences* 18(11): 878. By permission of the publisher (Taylor & Francis Ltd., http://www.tandf.co.uk/journals).

Mean weekly and session training volumes (distances) for sprint and middle-distance swimmers during the taper were reported by Stewart and Hopkins (2000). The weekly distance during the taper was significantly less than during previous training phases, and so was the session distance. These authors indicated that there were substantial reductions in all measures of training distance, easy swimming excepted, from the build-up phase to the end of the taper, for both sprinters and middle-distance specialists (figure 5.8).

The benefits of significant, progressive 50% to 90% reductions in training volume during the taper have been reported by several researchers in swimming, running, cycling, triathlon, and strength training (Houmard and Johns 1994, McNeely and Sandler 2007, Mujika 1998, Mujika and Padilla 2003a).

AT A GLANCE   **Reducing Load by Decreasing Volume**

There is enough scientific evidence to state confidently that training load reductions during the taper should be programmed at the expense of training volume. The body of research suggests that for most athletes, maximal performance gains are attained with training volume reductions of 41% to 60% of pretaper training, although performance benefits may also arise with smaller and bigger training volume reductions. These volume reductions usually affect all components of a training program.

Frequency

Hickson and Rosenkoetter (1981) provided evidence that it is possible for recently trained individuals to maintain the 20% to 25% gains in $\dot{V}O_2$max attained during 10 weeks of endurance training for at least 15 weeks of reduced training frequency, whether this reduction amounted to one third or two thirds (from 6 days/week to either 4 or 2 days/week) of previous values; figure 5.9). Similar results have been observed in strength-trained subjects (Graves et al. 1988). Several physiological and

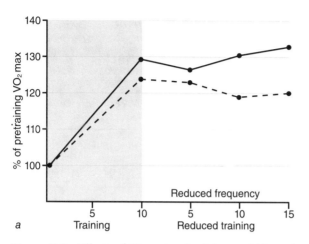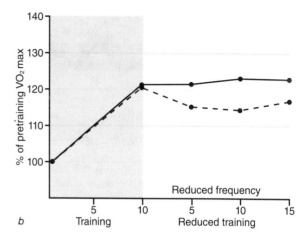

Figure 5.9 Effects of 10 weeks of training and 15 weeks of reduced training at one third (solid lines) and two-thirds (broken lines) reductions in frequency on *(a)* cycling and *(b)* treadmill running maximum oxygen uptake ($\dot{V}O_2$max).

Adapted, by permission, from R.C. Hickson, C. Foster, M.L. Pollock, et al., 1985, "Reduced training intensities and loss of aerobic power, endurance, and cardiac growth," *Journal of Applied Physiology* 58(Feb): 497. Permission conveyed through Copyright Clearance Center, Inc.

performance measures are retained or improved as a result of 2 to 4 weeks of reduced training frequencies in cyclists, runners, and swimmers (Mujika and Padilla 2003a). For instance, Johns and colleagues (1992) reported increased power and performance in competitive swimmers who reduced training frequency by 50% during 10 and 14 days of taper, and Dressendorfer and colleagues (2002b) observed a significant improvement in a 20 km cycling time trial simulation after a 50% reduction in training frequency during a 10-day taper.

The only available report that compared a high-frequency taper (maintenance of a daily training frequency) and a moderate-frequency taper (33% reduction in training frequency, i.e., resting every third day of the taper) in highly trained middle-distance runners concluded that training daily during a 6-day taper brought about significant 1.93% performance gains in an 800 m race, whereas resting every third day of the taper resulted in a nonsignificant gain of only 0.39%. Given that no differences in the physiological responses to the taper were found between groups, in the absence of systematic psychometric measurements before and after the taper, and in accordance with previous suggestions (Houmard and Johns 1994, Kubukeli et al. 2002, Neufer 1989), the authors attributed these results to a potential "loss of feel" during exercise (Mujika et al. 2002a).

According to Bosquet and colleagues (2007), decreasing training frequency has not been shown to significantly improve performance (tables 5.7 and 5.8). However, these authors point out that the decrease in training frequency often interacts with other training variables, particularly training volume and intensity, which makes it difficult to isolate the precise effect of a reduction in training frequency on performance.

Table 5.7   Effect of Decreasing or Not Decreasing Training Frequency During the Taper on Overall Effect Size for Taper-Induced Changes in Performance

Decrease in training frequency	Overall effect size, mean (95% confidence interval)	n	p
Yes	0.24 (–0.03, 0.52)	176	.08
No	0.35 (0.18, 0.51)	302	.0001

Data reprinted, by permission, from L. Bosquet, J. Montpetit, D. Arvisais, et al., 2007, "Effects of tapering on performance: A meta-analysis," *Medicine & Science in Sports & Exercise* 39: 1358-1365.

Table 5.8 Effect of Decreasing or Not Decreasing Training Frequency During the Taper on Overall Effect Size for Taper-Induced Changes in Swimming, Running, and Cycling Performance

Decrease in training frequency	SWIMMING			RUNNING			CYCLING		
	Mean (95% confidence interval)	n		Mean (95% confidence interval)	n		Mean (95% confidence interval)	n	
Yes	0.35 (–0.36, 1.05)	54		0.16 (–0.17, 0.49)	74		0.95 (–0.48, 2.38)	25	
No	0.30 (0.10, 0.50)*	195		0.53 (0.05, 1.01)**	36		0.55 (–0.05, 1.15)***	55	

*$p \le .01$; **$p \le .05$; ***$p \le .10$.

Data reprinted, by permission, from L. Bosquet, J. Montpetit, D. Arvisais, et al., 2007, "Effects of tapering on performance: A meta-analysis," *Medicine & Science in Sports & Exercise* 39: 1358-1365.

Table 5.9 Number of Weekly Training Sessions Prescribed by 24 Coaches for Sprint (50 and 100 m) and Middle-Distance (200 and 400 m) Swimmers for the Summer and Winter Seasons (Mean ± SD)

	BUILD-UP		SPECIALITY		TAPER		POSTCOMPETITION	
	Sprint	Middle-distance	Sprint	Middle-distance	Sprint	Middle-distance	Sprint	Middle-distance
Weekly sessions	6.8 ± 1.9	7.8 ± 2.0	7.3 ± 1.9	8.1 ± 1.7	6.6 ± 2.0	7.3 ± 2.0	5.0 ± 1.0	5.0 ± 1.2

Data reprinted from A.M. Stewart and W.G. Hopkins, 2000, "Seasonal training and performance of competitive swimmers," *Journal of Sports Sciences* 18(11): 877. By permission of the publisher (Taylor & Francis Ltd., http://www.tandf.co.uk/journals).

Stewart and Hopkins (2000) reported on the training frequency prescribed by 24 swimming coaches for sprint and middle-distance swimmers in each training phase of a periodized program. As can be seen in table 5.9, the number of weekly training sessions remained fairly constant throughout the season, with the exception of the postcompetition phase.

Taken together, the results discussed here suggest that whereas training adaptations can be readily maintained with quite low training frequencies in moderately trained individuals (30-50% of pretaper values), much higher training frequencies should be recommended for highly trained athletes (>80%), especially in the more technique-dependent sports such as swimming, rowing, skiing, and kayaking (McNeely and Sandler 2007, Mujika and Padilla 2003a).

AT A GLANCE Reducing Training Frequency

Studies evaluating the consequences of reduced training frequency during times of reduced training like the taper suggest that manipulations of this training variable affect moderately trained subjects and highly trained athletes differently. The former can readily maintain their physiological and performance adaptations with training frequencies representing just 30% to 50% of pretaper training, whereas highly trained athletes would be better off maintaining training frequencies similar to those used before the taper. Otherwise, there may be an increased risk of losing the much-needed "feel" for their sport, particularly in the more technique-dependent cyclic events.

Duration of the Taper

Assessing the most suitable duration of a taper for an individual athlete is one of the most difficult challenges for coaches and sport scientists. Positive physiological, psychological, and performance adaptations have been reported as a result of taper programs lasting 4 to 14 days in cyclists and triathletes, 6 to 7 days in middle- and long-distance runners, 10 days in strength trained athletes, and 10 to 35 days in swimmers (Mujika and Padilla 2003a). Unfortunately, the time frame that separates the benefits of a successful taper from the negative consequences of insufficient training (Mujika and Padilla 2000, Neufer 1989) has not been clearly established. Considering changes in blood lactate concentration and performance times derived from a test work set, Kenitzer (1998) concluded that a taper of approximately 2 weeks represented the limit of recovery and compensation time before detraining became evident in a group of female swimmers. Kubukeli and colleagues (2002) suggested that the optimum taper duration may be influenced by previous training intensity and volume, with athletes who trained harder and longer requiring roughly 2 weeks to fully recover from training while maximizing the benefits of training and those who reduced their amount of high-intensity training needing a shorter taper to prevent a loss of fitness.

Stewart and Hopkins (2000) reported on the duration of each training phase over a season of competitive swimming in sprinters and middle-distance swimmers and identified a slight difference between both groups: Sprinters tapered for almost 4 weeks, whereas middle-distance specialists tapered for less than 3 weeks. From this study, it emerged that the average swim coach prescribed a slightly longer taper for sprinters than for middle-distance swimmers (table 5.10), but no evidence is available to confirm this as an optimal approach to taper duration.

Bosquet and colleagues (2007) found a dose–response relationship between the duration of the taper and the performance improvement (tables 5.11 and 5.12). A

Table 5.10 Duration (Weeks) of Each Training Phase Prescribed by 24 Coaches for Sprint (50 and 100 m) and Middle-Distance (200 and 400 m) Swimmers for the Summer and Winter Seasons (Mean ± SD)

	BUILD-UP		SPECIALITY		TAPER		POSTCOMPETITION	
	Sprint	Middle-distance	Sprint	Middle-distance	Sprint	Middle-distance	Sprint	Middle-distance
Duration, weeks	12.1 ± 3.8	12.4 ± 4.5	7.0 ± 3.9	7.7 ± 4.3	3.8 ± 2.2	2.7 ± 1.4	3.1 ± 1.9	2.8 ± 1.9

Data reprinted from A.M. Stewart and W.G. Hopkins, 2000, "Seasonal training and performance of competitive swimmers," *Journal of Sports Sciences* 18(11): 877. By permission of the publisher (Taylor & Francis Ltd., http://www.tandf.co.uk/journals).

Table 5.11 Effect of Taper Duration on Overall Effect Size for Taper-Induced Changes in Performance

Duration of the taper	Overall effect size, mean (95% confidence interval)	n	p
≤7 days	0.17 (–0.05, 0.38)	164	.14
8-14 days	0.59 (0.26, 0.92)	176	.0005
15-21 days	0.28 (–0.02, 0.59)	84	.07
≥22 days	0.31 (–0.14, 0.75)	54	.18

Data reprinted, by permission, from L. Bosquet, J. Montpetit, D. Arvisais, et al., 2007, "Effects of tapering on performance: A meta-analysis," *Medicine & Science in Sports & Exercise* 39: 1358-1365.

Table 5.12 Effect of Taper Duration on Overall Effect Size for Taper-Induced Changes in Swimming, Running, and Cycling Performance

Duration of the taper	SWIMMING Mean (95% confidence interval)	n	RUNNING Mean (95% confidence interval)	n	CYCLING Mean (95% confidence interval)	n
≤7 days	−0.03 (−0.41, 0.35)	54	0.31 (−0.08, 0.70)	52	0.29 (−0.12, 0.70)	47
8-14 days	0.45 (−0.01, 0.90)***	84	0.58 (0.12, 1.05)*	38	1.59 (−0.01, 3.19)**	33
15-21 days	0.33 (0.00, 0.65)**	75	−0.08 (−0.95, 0.80)	10	No data available	
≥22 days	0.39 (−0.08, 0.86)	36	−0.72 (−1.63, 0.19)	10	No data available	

*p ≤ .01; **p ≤ .05; ***p ≤ .10.

Data reprinted, by permission, from L. Bosquet, J. Montpetit, D. Arvisais, et al., 2007, "Effects of tapering on performance: A meta-analysis," *Medicine & Science in Sports & Exercise* 39: 1358-1365.

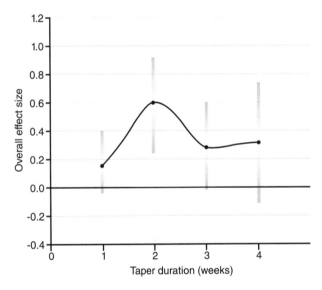

Figure 5.10 Dose–response curve for the effect of taper duration on performance.

Reprinted, by permission, from L. Bosquet, J. Montpetit, D. Arvisais, et al., 2007, "Effects of tapering on performance: A meta-analysis," *Medicine & Science in Sports & Exercise* 39: 1358-1365.

taper duration of 8 to 14 days was the borderline between the positive influence of fatigue withdrawal and the negative influence of losing adaptation (i.e., detraining).

Although performance improvements could also be expected as a result of tapers lasting 1, 3, or 4 weeks, negative results could also be experienced by some athletes, as indicated by the 95% confidence intervals reported in tables 5.11 and 5.12. The dose–response curve for the effect of taper duration on performance is shown in figure 5.10.

Some authors have used mathematical modeling in an attempt to optimize tapering strategies for individual athletes, including optimal taper duration (Fitz-Clarke et al. 1991, Morton et al. 1990, Mujika et al. 1996a, Mujika et al. 1996b). In one of these studies, the theoretical optimal taper duration in a group of national and international level swimmers was between mean values of 12 and 32 days, with a great intersubject variability (Mujika et al. 1996a), which leads to the conclusion that taper duration must be individually determined for athletes in accordance with their specific profiles of adaptation to training and loss of training-induced adaptations. Differences in the physiological and psychological adaptation responses to the training reduction and the use of an overload intervention, or lack thereof, in the weeks preceding the taper are some of the variables accounting for the reported interindividual variability in optimal taper duration (Bosquet et al. 2007).

Recent mathematical modeling simulations (Thomas et al. 2008) suggest that the training performed in the lead-up to the taper greatly influences the optimal individual duration of the taper. A 20% increase over normal training during 28 days before the taper requires a step reduction in training of around 65% during 3 weeks, instead of 2 weeks when no overload training is performed. However, a progressive taper requires a smaller reduction of training over a longer duration than a step taper, whatever the pretaper training. The impact of the pretaper training on the duration of the optimal taper seems obvious in regard to the reduction of the accumulated fatigue. Overload training before the taper causes a greater stress that needs longer to recover. Nevertheless, this impact could be also explained by the positive adaptations to training because higher training loads could make adaptations peak at a higher level but take longer to produce (Thomas et al. 2008) (see also the section titled Computer Simulations in chapter 7).

Millet and colleagues (2005) used mathematical modeling to describe the relationships between training loads and anxiety and perceived fatigue as a new method for assessing the effects of training on the psychological status of the athletes, in this case four professional triathletes. It was observed that the time for self-perceived fatigue to return to its baseline level was 15 days, which was close to the time modeled by previous researchers as optimal for tapering (Busso et al. 1994, Busso et al. 2002, Fitz-Clarke et al. 1991). The authors of the investigation concluded that the use of a simple questionnaire to assess anxiety and perceived fatigue could be used to adjust the optimal duration of tapering (Millet et al. 2005).

Determining Taper Duration

AT A GLANCE

Determining the duration of a taper is not easy. Available scientific reports show beneficial physiological and performance effects after tapers lasting as little as 4 days and as long as 5 weeks. No study has shown that the duration or intensity of the event an athlete is preparing for determines taper duration. According to recent investigations, overreaching an athlete before the taper would require not a bigger percentage training reduction during the taper but rather a longer taper. Most athletes will benefit from a 2-week taper, but shorter or longer tapers may be optimal for some athletes, depending on their individual profiles of fitness loss and fatigue dissipation. These individual profiles can be made using simple questionnaires to assess anxiety and perceived fatigue.

Environmental Factors and Tapering

Athletes preparing for international competition often have to deal with environmental factors that can negatively affect their performance capabilities. Long-distance travel across time zones and challenging environments characterized, for example, by high temperatures or reduced oxygen content in the air may significantly disrupt an athlete's comfort, requiring specific physiological adaptations. Training and tapering programs in the lead-up to important events taking place in such environments must also be specifically adapted.

Researchers from Australia, the Basque Country, and Great Britain have recently addressed the interactions between tapering, performance peaking, and environmental factors that international-caliber athletes must often deal with (Pyne et al. 2009).

Travel Across Time Zones, Peaking, and Chronobiology

Olympic athletes are frequently required to travel across continental boundaries for purposes of training or competing. Long-haul flights lead to travel fatigue, a relatively transient malaise that is quickly overcome with rest and sleep. Crossing multiple meridians causes desynchronization of human circadian rhythms and leads to the syndrome known as jet lag. This condition can persist for some days, depending on the number of time zones crossed, the direction of flight, the times of departure and arrival, and individual factors (Waterhouse et al. 2007b). Concomitant with the experience of jet lag, a range of performance measures are impaired until the endogenous "body clock," located in the hypothalamus, is retuned to the new local time. Decrements have been reported in muscle strength, reaction times, and subjective states indicative of arousal (Reilly et al. 2001).

Among the recommendations to help cope with jet lag during the time of desynchronization is lowering the training load (Reilly et al. 2007a). This reduction would compensate for the temporary decrease in physical capabilities associated with jet lag and avoid exposing the athlete to risk of injury in training (especially if complex practices are attempted) when symptoms are experienced. In effect, a temporary taper is imposed by the need to cope with the malaise of jet lag until readjustment has occurred, the duration of which depends mainly on the direction of flight and the number of time zone transitions. The jet-lag–imposed modifications to the training load should be considered in context of the overall training program and the taper leading to competition.

The amount of time that an athlete will require to adjust the body clock can be incorporated into the taper when competition requires travel across multiple meridians. It is logical that sufficient time be allowed for the athlete to adjust completely to the new time zone before competing (Waterhouse et al. 2007b). The readjustment might constitute a part of the lowered training volume integral to the taper. Allowance should be made for the timing of training over the first few days because training in the morning is not advocated after an athlete has traveled eastward so that a phase delay rather than the desired phase advance is not erroneously promoted (Reilly et al. 2005). There also seems little point in training hard at home prior to embarkation, because arriving tired at the airport of departure may slow adjustment later (Waterhouse et al. 2003). Similarly, attempting to shift the phase of the body clock in the required direction for some days prior to departure is counterproductive, because performance (and hence training quality) may be disrupted by this strategy (Reilly and Maskell 1989).

Tapering should proceed as planned in the company of jet lag, even if the interactions between body clock disturbances and the recovery processes associated with tapering have not been fully delineated. These interactions might implicate sleep, digestive, and immune functions. Although quality of sleep is an essential component of recovery processes, napping at an inappropriate time of day when adjusting to a new time zone may delay resynchronization (Minors and Waterhouse 1981); however, in certain circumstances, a short nap of about 30 min can be restorative (Waterhouse et al. 2007a). Suppression of immune responses is more likely to be linked with sleep disruption than with jet lag per se (Reilly and Edwards 2007).

The circadian rhythm in digestion is largely exogenous, and jet lag is associated mainly with a displacement of appetite rather than reduced energy intake (Reilly et al. 2007b). Therefore, a behavioral approach to readjustment of the body clock may be in line with the moderations of training associated with a taper. Athletes, coaches, managers, and support staff should implement strategies to minimize the effects of travel stress prior to departure, during long-haul international travel, and upon arrival at the destination.

Heat and Altitude

A taper is dependent on removal or minimization of the athlete's habitual stressors, thereby permitting physiological systems to replenish their capabilities or even undergo supercompensation. There is very little scientific information with respect to possible interactions with environmental variables on tapering processes in athletes, whether the stressor is heat, cold, or altitude. Experimental work on the additive effects of altitude on climatic stress and travel fatigue or jet lag is lacking (Armstrong 2006). This gap in knowledge is largely attributable to the enormous difficulties in addressing these problems adequately in experimental designs and the challenges that researchers in the field face in regard to controlling all the variables involved. Nevertheless, the likely effects of environmental factors must be considered in a systematic way when tapering is prescribed within the athlete's or team's annual plan.

The use of training camps, whether for altitude training or warm-weather training to secure heat acclimatization, poses a different set of quandaries than those presented by travel schedules. Both altitude and hot environments present hostile conditions for athletes whose primary objective for the visit is to subject themselves to the prevailing stressors. The principle is that physiological benefits derived from the adaptations to the new environment presumably transfer during the taper to enhance competitive performance. In both conditions, the absolute exercise intensity is lowered by necessity, even if the relative physiological stress is the same as that usually encountered. This uncoupling of physiological and biomechanical loading may have unknown consequences in terrestrial sports for the mechanisms associated with the tapering response, particularly when the exercise intensity is light to moderate during the first tentative exposures to heat or altitude.

Tapering in hot conditions prior to competition is compatible with the reduction in training volume advocated when encountering heat stress. The increased glycogen utilization associated with exercise in the heat should be compensated by the reduced training load, both intensity and duration (Armstrong 2006). Athletes should be acclimatized to heat; otherwise performance in the forthcoming competition might be compromised. In winter-sport events there is not a corresponding problem, because protection against climatic conditions and the initial diuresis that occurs is secured by behavioral measures, such as donning appropriate clothing, seeking shelter when necessary, and restoring hydration status.

At altitude, maximal oxygen uptake is reduced according to the prevailing ambient pressure. An immediate consequence is that the exercise intensity or power output at a given relative aerobic loading is decreased. In the first few days at altitude, a respiratory alkalosis occurs attributable to the increased ventilatory response to hypoxic conditions. This condition is normally self-limiting because of a gradual renal compensation. Athletes using training camps at altitude resorts recognize that a reduction in training load is imperative at altitude, prior to an increase as the initial phase of acclimatization occurs. The extra hydration requirements attributable to the dry ambient air and the initial diuresis, combined with plasma volume changes (Rusko et al. 2004), increased utilization of carbohydrate as substrate for exercise (Butterfield et al. 1992), and propensity to sleep apnea (Pedlar et al. 2005), run counter to the generally supportive environment associated with tapering. In this instance, the reduced training load would not substitute for a taper. There is the added risk of illness caused by decreased immunoreactivity associated with exposure to altitude (Rusko et al. 2004). Maximal cardiac output may also be reduced in the course of a typical 14- to 21-day sojourn to altitude as a result of the impairment in training quality. Altitude training camps should therefore be lodged strategically in the annual plan to avoid unwanted, if unknown, interactions with environmental variables.

Altitude training is used in many sports at the elite level for conditioning purposes. It is accepted as good practice among elite swimmers and rowing squads, for example, despite absence of compelling evidence of its effectiveness. Apart from its necessity in preparing for competition at altitude, altitude training is accepted by practitioners as enhancing performance at sea level. The optimal timing of the return to sea level is not known and is an issue relatively neglected by researchers in the field, with a few exceptions (Ingjer and Myhre 1992). Athletes have been advised to reduce training for a time before competing after altitude training, which constitutes a form of tapering. The extent of the benefit, as well as its variation between individuals, has not been adequately explored.

AT A GLANCE   **Dealing With Environmental Factors and the Taper**

Environmental factors like travel across time zones, heat, and altitude may interfere with an athlete's preparation for international-level competition. Crossing multiple meridians causes desynchronization of the body clock and leads to the syndrome known as jet lag, which impairs muscle strength, reaction times, and level of arousal. Reducing the training load has been recommended as a means to cope with jet lag, and this training reduction should be integrated in an athlete's taper program.

The interactions between taper and environmental stressors like heat, cold, and altitude have not been studied. Physiological benefits derived from the adaptations to the new environment presumably transfer during the taper to enhance competitive performance. Tapering in hot conditions prior to competition seems to be compatible with the reduction in training volume advocated when encountering heat stress. Similarly, training camps at altitude usually require an initial reduction in training load, which in itself may constitute a form of tapering.

Chapter Summary

Maintaining training intensity (i.e., "quality training") is absolutely necessary to retain and enhance training-induced adaptations during tapering, but it is obvious that reductions in the other training variables should allow for sufficient recovery to optimize performance. As we have seen, reductions in training volume appear to induce positive physiological, psychological, and performance responses in highly trained athletes. According to the most recent analysis of the scientific literature, a safe bet in terms of training volume reduction would be 41% to 60%, but performance benefits could be attained with somewhat smaller or bigger volume reductions.

High training frequencies (>80%) seem to be necessary to avoid detraining and "loss of feel" in highly trained athletes. Conversely, training-induced adaptations can be readily maintained with very low training frequencies in moderately trained individuals (30-50%).

The optimal duration of the taper is not known. Indeed, positive physiological and performance adaptations can be expected as a result of tapers lasting 4 to 28 days, yet the negative effects of complete inactivity are readily apparent in athletes. When we are unsure about the individual adaptation profile of a particular athlete that would determine his or her optimal taper duration, 2 weeks seems to be the suit-all taper duration. On the basis of their own experience, McNeely and Sandler (2007) have developed some guidelines for the recommended duration of a taper based on the number of weekly training hours. These guidelines, requiring scientific confirmation, are provided in table 5.13.

Table 5.13   Training Time for Major Taper

Training hours/week	Major taper, days
6-10	7
11-15	14
>15	21-30

Reprinted, by permission, from E. McNeely and D. Sandler, 2007, "Tapering for endurance athletes," *Strength and Conditioning Journal* 29(5): 21.

When athletes are preparing for international competition, environmental factors like travel across time zones, heat, and altitude may interfere with the taper. Crossing multiple meridians causes desynchronization of the body clock, leading to jet lag and impairing muscle strength, reaction times, and level of arousal. A training load reduction can help the athlete cope with jet lag, and this training reduction should be integrated into the taper program.

The interactions between taper and environmental stressors like heat, cold, and altitude are yet to be studied. The physiological benefits of adapting to the stressful environment should transfer during the taper to enhance competitive performance. Tapering in hot environments before competition seems to be compatible with the reduction in training volume recommended when facing heat stress. Altitude training camps also require an initial reduction in training load, which may in itself constitute a form of tapering.

chapter 6

Designing the Taper for Performance Improvements

The final and major goal of a taper is to optimize competition performance. Most studies dealing with progressive tapers in athletes have reported significant performance improvements in various sports including swimming, running, cycling, rowing, and triathlon. Some investigators have determined performance changes in actual competition, whereas others have reported on laboratory or field-based criterion performance measures. In this chapter, the size of these performance improvements, the factors potentially affecting them, and their meaningfulness in terms of competition outcome are addressed.

Observed Performance Gains in Various Sports

Tapering-induced performance gains, which have variously been attributed to increased levels of muscular force and power; improvements in neuromuscular, hematological, and hormonal function; and psychological status of the athletes (see chapters 2-4), are usually in the range of 0.5% to 6.0% for competition performance measures but can reach 22% in noncompetitive criterion measures (table 6.1). It is important to establish the validity of the performance tests and their relationship with actual performance in a specific competition event (see section titled Meaningfulness of Tapering-Induced Performance Gains).

In an observational investigation on Olympic swimmers, Mujika and colleagues (2002b) reported on performance change in 99 individual swimming events during the final 3 weeks of training (generally coincident with the taper) in the lead-up to the Sydney 2000 Olympic Games. The overall performance change during the 3 weeks prior to competition for all swimmers was 2.18% ± 1.50%, with a maximum loss of –1.14% and a maximum gain of 6.02%. A total of 91 of the 99 analyzed performances were faster and only 8 performances were slower after the taper. Performance improvements were not significantly different among events, and they ranged from 0.64 ± 1.48% in 400 m freestyle (event with the smallest mean improvement) to 2.96 ± 1.08% in 200 m butterfly (the event with the largest mean improvement) (table 6.2).

The 2.6% performance improvement attained during the taper by the male swimmers in the study by Mujika and colleagues (2002b) was somewhat lower than some

Table 6.1 Effects of the Taper on Performance in Trained Athletes

Study and year	Athletes	Taper duration, days	Performance measure	Performance outcome, %
Costill et al. 1985	Swimmers	14	50- to 1,650-yard (46-1,509 m) competition	2.2-4.6 ↑
Cavanaugh & Musch 1989	Swimmers	28	50- to 1,650-yard (46-1,509 m) competition	2.0-3.8 ↑
Houmard et al. 1989	Runners	10	Incremental maximal test	Unchanged
Houmard et al. 1990a	Runners	21	5K indoor race	Unchanged
Costill et al. 1991	Swimmers	14-21	Competition	≈3.2 ↑
D'Acquisto et al. 1992	Swimmers	14-28	100, 400 m time trial	4.0-8.0 ↑
Jeukendrup et al. 1992	Cyclists	14	8.5 km outdoor time trial	7.2 ↑
Johns et al. 1992	Swimmers	10-14	50- to 400-yard (46-366 m) competition	2.0-3.7 ↑
Shepley et al. 1992	Runners	7	Treadmill time to exhaustion	6-22 ↑
McConell et al. 1993	Runners	28	5K indoor race	1.2 ↓
Flynn et al. 1994	Runners Swimmers	21	Treadmill time to exhaustion 25-yard (23 m), 400-yard (366 m) time trial	Unchanged ≈3 ↑
Gibala et al. 1994	Strength-trained athletes	10	Voluntary elbow flexor strength	≈7 ↑
Houmard et al. 1994	Runners	7	5K treadmill time trial	2.8 ↑
Martin et al. 1994	Cyclists	14	Incremental maximal test	8.0 ↑
Zarkadas et al. 1995	Triathletes	14	5K field time-trial run Incremental maximal test	1.2-6.3 ↑ 1.5-7.9 ↑
Mujika et al. 1996b	Swimmers	28	100-200 m competition	0.4-4.9 ↑
Raglin et al. 1996	Swimmers	28-35	Competition	2.0 ↑
Stone et al. 1996	Weightlifters	7-28	Competition	8.0-17.5 kg ↑
Taylor et al. 1997	Swimmers	Not reported	Competition	1.3 ↑
Hooper et al. 1998	Swimmers	14	100, 400 m time trial	Unchanged
Kenitzer et al. 1998	Swimmers	14-28	4 × 100-yard (91 m) sub-maximal set	≈4 ↑

Study and year	Athletes	Taper duration, days	Performance measure	Performance outcome, %
Berger et al. 1999	Cyclists	14	4 km simulated pursuit	2.0 ↑
Hooper et al. 1999	Swimmers	14	100 m time trial	Unchanged
Bonifazi et al. 2000	Swimmers	14-21	100-400 m competition	1.5-2.1 ↑
Child et al. 2000	Runners	7	Simulated half-marathon	Unchanged
Martin & Andersen 2000	Cyclists	7	Incremental maximal test	≈6 ↑
Mujika et al. 2000	Runners	6	800 m competition	Unchanged
Smith 2000	Rowers	7	500 m simulated time trial	Unchanged
Steinacker et al. 2000	Rowers	7	2,000 m time-trial competition	6.3 ↑
Trappe et al. 2001	Swimmers	21	Competition	3.0-4.7 ↑
Rietjens et al. 2001	Cyclists	21	Incremental maximal test	Unchanged
Dressendorfer et al. 2002a, 2002b	Cyclists	10	20 km simulated time trial	1.2 ↑
Eliakim et al. 2002	Handball players	14	4 × 20 m sprint, vertical jump	2.1-3.2 ↑
Mujika et al. 2002a	Runners	6	800 m competition	0.4-1.9 ↑
Maestu et al. 2003	Rowers	14	2,000 m ergometer time trial	Unchanged
Margaritis et al. 2003	Triathletes	14	30 km outdoor duathlon	1.6-3.6 ↑
Neary et al. 2003a	Cyclists	7	20 km simulated time trial	5.4 ↑
Neary et al. 2003b	Cyclists	7	40 km simulated time trial	2.2-4.3 ↑
Harber et al. 2004	Runners	28	8 km outdoor race	1.1 ↑
Neary et al. 2005	Cyclists	7	20 km simulated time trial	4.5 ↑
Trinity et al. 2006	Swimmers	21	50-1,500 m competition	4.5 ↑
Vollaard et al. 2006	Triathletes	7	15 min simulated time trial	4.9 ↑
Coutts et al. 2007a	Triathletes	14	3 km run time trial	3.9 ↑
Izquierdo et al. 2007	Strength-trained athletes	28	1 RM strength	2.0-3.0 ↑
Papoti et al. 2007	Swimmers	11	200 m time trial	1.6 ↑

↑ indicates improvement; ↓ indicates decline.

Table 6.2 Performance Times (s) Before and After the Final 3 Weeks of Training (F3T)

Event	MALE				FEMALE				TOTAL	
	n	Pre-F3T	Post-F3T	% change	n	Pre-F3T	Post-F3T	% change	n	Total % change
50 Freestyle	4	22.89 ± 0.40	22.49 ± 0.33	1.73 ± 1.60	6	26.87 ± 0.74	26.33 ± 1.01*	2.06 ± 1.53	10	1.93 ± 1.48
100 Freestyle	8	52.01 ± 1.65	50.66 ± 1.72**	2.59 ± 1.49	5	57.94 ± 1.82	56.58 ± 1.42**	2.33 ± 0.69	13	2.49 ± 1.21
200 Freestyle	6	114.83 ± 5.57	111.05 ± 4.20**	3.25 ± 1.56	4	121.96 ± 0.78	120.14 ± 1.38	1.49 ± 1.13	10	2.55 ± 1.61
400 Freestyle	2	239.17 ± 11.35	234.78 ± 9.28	1.82 ± 0.78	2	255.44 ± 0.00	256.80 ± 1.78	−0.53 ± 0.70	4	0.64 ± 1.48
800 Freestyle					3	525.22 ± 1.84	519.64 ± 3.23	1.06 ± 0.89	3	1.06 ± 0.89
100 Backstroke	4	57.89 ± 3.42	56.20 ± 2.11	2.82 ± 2.18	4	63.34 ± 0.84	62.63 ± 0.75	1.09 ± 2.06	8	1.96 ± 2.17
200 Backstroke	2	121.09 ± 2.19	118.60 ± 1.43	2.05 ± 0.59	4	136.39 ± 2.68	134.81 ± 1.70	1.15 ± 0.77	6	1.45 ± 0.80
100 Breaststroke	5	67.17 ± 5.19	65.24 ± 4.66*	2.84 ± 1.73	5	73.93 ± 6.70	72.50 ± 6.49*	1.92 ± 1.26	10	2.38 ± 1.51
200 Breaststroke	3	136.47 ± 2.54	133.08 ± 0.71	2.45 ± 2.35	3	148.78 ± 0.98	147.13 ± 1.41	1.11 ± 1.59	6	1.78 ± 1.94
100 Butterfly	6	55.45 ± 1.65	54.24 ± 1.85**	2.19 ± 1.01	5	62.61 ± 3.24	60.96 ± 3.41*	2.64 ± 2.06	11	2.39 ± 1.50
200 Butterfly	5	122.13 ± 2.12	118.59 ± 1.91***	2.90 ± 0.58	4	132.71 ± 2.76	128.65 ± 2.10*	3.04 ± 1.63	9	2.96 ± 1.08
200 Individual medley	4	127.02 ± 2.81	122.92 ± 0.74*	3.20 ± 1.63	2	140.76 ± 1.86	137.59 ± 2.62	2.25 ± 0.57	6	2.88 ± 1.38
400 Individual medley	1	261.40	260.31	0.42	2	288.91 ± 3.08	283.77 ± 3.19	1.78 ± 0.06	3	1.33 ± 0.79

Values are mean ± standard deviation. Significantly different from pre-F3T: *p < .05; **p < .01; and ***p < .001.

Reprinted, by permission, from I. Mujika, S. Padilla, and D. Pyne, 2002, "Swimming performance changes during the final 3 weeks of training leading to the Sydney 2000 Olympic Games," *International Journal of Sports Medicine* 23: 585.

values previously published in the tapering literature. For example, Costill and colleagues (1985) reported a mean performance improvement of 3.1% as a result of a 2-week taper in a group of 17 collegiate male swimmers. Studying a similar population of 24 college male swimmers tapering for 2 to 3 weeks, the same group of researchers observed a 3.2% gain in performance (Costill et al. 1991). Johns and colleagues (1992) also reported an average performance improvement of 2.8% ± 0.3% with a 10- to 14-day taper.

Performance improvement differences between the studies mentioned in the previous paragraph and the study by Mujika and colleagues (2002b) could be partly attributable to the higher performance level of the swimmers analyzed in the latter investigation. Indeed, some of the highest mean performance gains with the taper (7.96% and 5.00% in 100 m and 400 m, respectively) have been reported in high school swimmers (D'Acquisto et al. 1992), whereas values of 2.6% (Cavanaugh and Musch 1989) and 2.32% ± 1.69% (Mujika et al. 1996b) have been observed in national- and international-level male swimmers during tapers lasting 4 weeks. Bonifazi and colleagues (2000) analyzed the effects of a taper lasting 2 to 3 weeks in international-level male swimmers during two consecutive seasons. Performance improved by 1.48% during the first season and 2.07% during the second. Taken together, the results of these investigations indicate that the taper in international-level swimmers usually produces an average improvement in performance in the range of 1.5% to 2.5%.

Figure 6.1 shows percentage performance improvement with the taper (final 3 weeks of training) in males and females participating in events of varying distance. No significant difference was observed between any of the analyzed swimming distances, which suggests that the metabolic contribution to energy provision during competition, which varies with racing distance, does not affect the potential gain that can be obtained during a 3-week taper. This suggestion is further supported by the data reported in table 6.1, in which similar performance improvements can be observed in different sports with criterion performance measures ranging in duration from a few seconds (e.g., 1RM strength or a 25-yard [23 m] maximal swim) to about 1 hr (e.g., 40 km cycling time trial).

Similarly, the comparison between freestyle and form (backstroke, breaststroke, butterfly, and individual medley) events showed no differences in the magnitude of performance improvement (figure 6.2), suggesting that technical and biomechanical aspects of competition do not necessarily affect the performance outcome of a taper. In addition, performance change with the taper was not significantly different among swimmers from the 14 different countries represented in the sample, ranging from 0.13% ± 0.28% for Nigerian swimmers to 3.98% ± 2.18% for swimmers representing Swaziland.

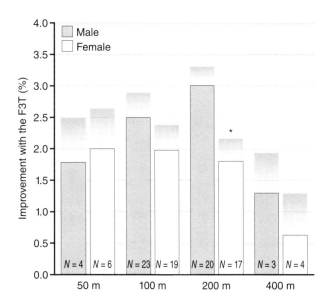

Figure 6.1 Percentage performance improvement with the final 3 weeks of training (F3T) in male and female 50 m (freestyle only), 100 m (freestyle, backstroke, breaststroke, and butterfly), 200 m (freestyle, backstroke, breaststroke, butterfly, and individual medley), and 400 m (freestyle and individual medley) events. *Significant difference (p < .05) between males and females. Values are mean ± SE.

Reprinted, by permission, from I. Mujika, S. Padilla, and D. Pyne, 2002, "Swimming performance changes during the final 3 weeks of training leading to the Sydney 2000 Olympic Games," *International Journal of Sports Medicine* 23: 584.

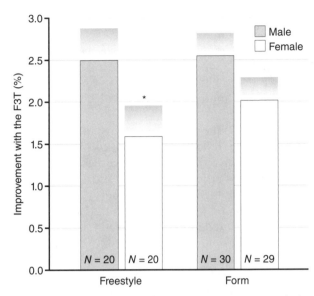

Figure 6.2 Percentage performance improvement with the final 3 weeks of training (F3T) in male and female freestyle and form (backstroke, breaststroke, butterfly, and individual medley) events. *Significant difference ($p < .05$) between males and females. Values are mean ± SE.

Reprinted, by permission, from I. Mujika, S. Padilla, and D. Pyne, 2002, "Swimming performance changes during the final 3 weeks of training leading to the Sydney 2000 Olympic Games," *International Journal of Sports Medicine* 23: 584.

In swimming, one of the most thoroughly studied sports with regard to the effects of tapering, performance improvements regardless of the specific event have been reported by several authors. In male swimmers, Costill and colleagues (1985) observed gains ranging from 2.2% in 100-yard (91 m) freestyle ($n = 2$) and 200-yard (182 m) butterfly ($n = 3$) and 4.6% in 200-yard (182 m) individual medley ($n = 5$). Johns and colleagues (1992) reported a minimum gain of 2.0% in 100-yard (91 m) breaststroke ($n = 1$) and a maximum of 3.7% in 100-yard (91 m) ($n = 5$) and 200-yard (182 m) freestyle ($n = 2$). Similar values, between 2.0% ($n = 6$) and 3.8% ($n = 4$), were reported by Cavanaugh and Musch (1989) for 50-yard (46 m) freestyle and 200-yard (182 m) breaststroke, respectively. All of these were quite similar to the 1.73% and 3.25%,

respectively, observed in 50 m ($n = 4$) and 200 m freestyle ($n = 6$) for males in the study by Mujika and colleagues (2002b). Once again, these results support the notion that tapering elicits a relatively consistent improvement across different competitive swimming events.

AT A GLANCE ## Using the Data to Set Goals

The data reported on performance change across sports, distances, and events provide a quantitative framework for coaches and athletes to set realistic performance goals based on individual performance levels before the tapering phase leading to important competitions, as illustrated in the following examples.

- An elite 100 m breaststroker races before tapering and his time is 62.00 s. Chances are that after an efficient taper, his competition performance will improve by about 3%, for a swim time of 60.14 s. If the taper program is not particularly efficient, he may improve by just 0.5%, for a time of 61.69 s. On the other hand, a particularly efficient taper inducing a gain of 6% will help this swimmer clock a world record–smashing time of 58.28 s!

- A highly trained 5K runner's time at a minor competition prior to the taper is 14 min 10 s. Expecting him to run 12 min 45 s after the taper would be an unrealistic goal, because this time represents a rather unlikely 10% improvement. More likely expectations for this runner would be between 14 min 6 s (0.5% gain) and 13 min 19 s (6% gain).

Individual Differences

As mentioned in the previous section, eight of the swimmers in the study by Mujika and colleagues (2002b) did not improve their performance during the Olympic Games compared with the competition that took place 3 weeks earlier. As can be observed in table 6.1, unchanged performances have been reported during the taper in various sports and performance measures. A lack of performance gain during the taper could be the result of a poorly planned tapering strategy, an individually determined lack of physiological or psychological response to the reduced training, or a combination of both. Unfortunately, the specific effects of these potential influencing factors cannot be ascertained in most cases, because of either an insufficiently thorough description of the tapering strategy used by the researchers, the lack of individual response data in favor of group statistics, or the observational nature of a study design.

For instance, Mujika and colleagues (2002b) discussed that the observational nature of their investigation imposed several limitations on the interpretation of results and inferences that can be applied to the population of competitive swimmers from their study sample. No systematic or consistent biological or psychometric measurements were undertaken at the time of the two analyzed competitions (i.e., pre- and posttaper). Therefore, all discussed mechanisms possibly contributing to the observed performance improvements were rather speculative, although they were considered in the light of data reported in previously published swimming tapering studies.

The potential physiological and psychological mechanisms underpinning the variable performance changes associated with the taper are thoroughly described in chapters 2 through 4. Nevertheless, researchers from the Department of Biological Sciences of the University of Essex and the School of Life Sciences of Heriot-Watt University in the United Kingdom recently expressed an interesting view concerning the mechanisms of the commonly observed performance improvement after tapering (Vollaard et al. 2006), which could partly account for the individual differences in the response to a taper. These authors rightly point out that regardless of study design, studies investigating the effects of training manipulations on exercise performance will inevitably suffer methodological difficulties arising from the impossibility of blinding the subjects to the treatment. Given that most well-trained athletes are aware of the potential beneficial effects of a taper and that they will know whether their training load has been increased or decreased, a placebo effect cannot be discounted to account for at least part of the performance gain observed in tapering investigations (Vollaard et al. 2006). According to these authors, expectancy effects could provide a credible mechanism for the observed beneficial effects of tapering. This possibility, of course, deserves to be investigated (Vollaard et al. 2006), and if it is confirmed, it is conceivable that the expectancy effects could also be closely associated with individual variability in performance outcomes.

Assessing Individual Adaptation to the Taper   AT A GLANCE

As we have described in previous chapters, a variety of physiological (e.g., haptoglobin, reticulocytes, testosterone, cortisol, blood lactate, creatine kinase), and psychological (e.g., Profile of Mood States, Recovery–Stress Questionnaire for Athletes, rating of perceived exertion) markers have been shown to change during the taper, and because these changes sometimes correlate with performance variations, the implication is that they could help gauge the effects of a taper. Most of these markers, however, are too invasive, too cumbersome, too expensive, or not sufficiently specific. In my experience, there are two specific, noninvasive, relatively simple, and quite inexpensive tools that can provide extremely valuable information about an athlete's adaptation to the taper: one is called a chronometer (or stopwatch) and the other one verbal communication.

Sex Effect

Systematic studies on the possible existence of a sex effect regarding the performance consequences of tapering are lacking, but some reports have provided separate data on male and female athletes. In the study by Mujika and colleagues (2002b), the mean percentage improvement with the pre-Olympic taper was significantly higher in male (2.57% ± 1.45%) than in female (1.78% ± 1.45%) swimmers. In addition, males improved more than females (2.91% ± 1.38% vs. 1.80% ± 1.35%, $p < .05$) in the 200 m events (figure 6.1). Comparisons between freestyle and form (backstroke, breaststroke, butterfly, and individual medley) events showed that males improved more than females in the freestyle events (2.54% ± 1.50% vs. 1.60% ± 1.33%, $p < 0.05$) (figure 6.2). In a report concerning female swimmers, Mujika and colleagues (1996a) indicated a performance gain range between 0.47% in 100 m breaststroke ($n = 1$) and 5.42% in 100 m butterfly ($n = 1$), in contrast with the performance loss of 0.53% observed in 400 m freestyle ($n = 2$) and a maximum gain of 3.04% in 200 m butterfly ($n = 4$) observed by Mujika and colleagues (2002b).

Given that the investigation by Mujika and colleagues (2002b) was purely observational and that the influence of factors such as motivation, shaving, use of different swimming suits, or diet could not be controlled for, there was no apparent explanation for the significantly smaller mean performance improvement achieved by the female swimmers, especially in the absence of any accompanying biological or psychometric measurements.

In contrast with the observations by Mujika and colleagues (2002b), in a study on national- and international-level swimmers, performance improvements of 2.58% ± 1.96%, 2.95% ± 0.93%, and 2.56% ± 2.19% were reported for the females, consecutive to each of the three tapers carried out during a competitive season, lasting respectively 3, 4, and 6 weeks. These values were not significantly different from those of 3.12% ± 1.15%, 3.42% ± 2.19%, and 1.60% ± 0.92% attained by the males (Mujika et al. 1996a). Hooper and colleagues consistently observed small nonsignificant performance gains of 0.1% to 0.7% in international-level male and female swimmers, but separate values for each gender were not reported (Hooper et al. 1993, Hooper et al. 1998, Hooper et al. 1999). Similarly, Smith (2000) did not report any sex difference in the amplitude of performance improvement after a 1-week taper in 6 female and 10 male elite rowers.

Nevertheless, a sex effect has been suggested for the optimum taper duration. Indeed, Kenitzer (1998) recommended for female swimmers a taper duration of 2 weeks, because performance indexes began to deteriorate with longer tapers. Moreover, different researchers have put forward the suggestion that the relatively small 1.3% performance gains attained by female swimmers during a taper could be related to a deterioration of their mood state (Taylor et al. 1997). The preceding results, however, are in contrast with those of Mujika and colleagues (1996a), who did not observe sex differences in the optimal duration of the taper as determined by mathematical modeling, and with those of Morgan and colleagues (1987) and Raglin and colleagues (1991), who reported similar changes in perception of effort, global mood score, and total mood disturbance in male and female swimmers during tapering. More investigations are required to shed additional light on the interactions among tapering, sex, mood state, and competitive athletic performance.

AT A GLANCE   **Using the Data to Take Sex Difference Into Account**

The available research regarding the existence of a difference between males and females in the effects of the taper is inconclusive: Whereas some investigations have found bigger performance gains in males, the majority of studies have seen no difference between sexes; whereas some researchers suggest that the optimal taper duration should be no longer than 2 weeks in females, modeling studies report no differences with the optimal duration for males; whereas some investigators suggest a deterioration of mood states in females during the taper, others describe identical changes in males and females. There is no evidence to suggest that tapering programs should be differentially designed for males and females or that a sex effect exists concerning the adaptations and taper effects on performance.

Meaningfulness of Taper-Induced Performance Gains

The percentage performance changes reported in table 6.1 seem to be rather small, and some studies in the scientific literature report that taper-induced changes were not statistically significant. Although taper-induced improvements in performance may often be small and nonsignificant from a statistical point of view, they could be worthwhile in practical terms relative to competition results (Hopkins et al. 1999, Mujika et al. 2002b). Improvements of approximately one half of the typical within-subject variation in competition performance substantially increase a top athlete's chance of winning a medal. The magnitude of such practical improvements is 0.9% in track sprinters (Hopkins et al. 1999), 1.4% for highly trained swimmers (Stewart and Hopkins 2000), 1.5% for cross country runners, and up to 3% for marathon runners (Hopkins and Hewson 2001). Another important consideration is that laboratory or field-based performance tests are much less reliable measures of performance than competition itself, and enhancements in these tests may not necessarily translate into similar performance gains in competition (Hopkins et al. 1999).

In the study by Mujika and colleagues (2002b), percentage performance improvement with the taper was similar in Olympic medal winners, finalists, semifinalists, and qualifying heat swimmers. This was true for males, for females, and for all swimmers taken as a single group. However, among those athletes who swam only the qualifying heats, performance improvement with the taper was bigger in males (2.84% ± 1.67%) than in females (1.23% ± 1.44%) (figure 6.3). But this is the important part: These authors also calculated the percentage difference in performance time from first to fourth place (gold medal winner to first swimmer out of the medals) and from third to eighth place (bronze medal winner to last finalist) for all male and female events of the Olympic swimming finals. The percentage difference in performance time between first and fourth place was 1.62% ± 0.80% overall (males 1.48% ± 0.67%, females 1.77% ± 0.90%) and between third and eighth place 2.02% ± 0.81% overall (males 1.69% ± 0.56%, females 2.35% ± 0.91%).

In the study by Mujika and colleagues (2002b), the observed swim time improvement with the taper was considered very worthwhile in performance terms, because the differences between the gold medalist and the first swimmer out of the medals and between the bronze medal winner and the last swimmer in the final at the Olympic events were smaller than the mean improvement in swim time obtained during the taper. In other words, a very successful taper could take an athlete from finishing last in the Olympic final to winning a bronze medal or from not winning an Olympic medal to becoming an Olympic champion! This, of course, is true as long as the other finalists had a less successful tapering strategy.

Bosquet and colleagues (2007) indicated that because performance is a complex system whose whole is more than the

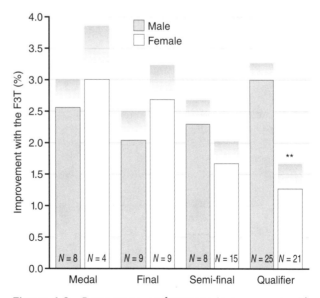

Figure 6.3 Percentage performance improvement with the final 3 weeks of training (F3T) in male and female Olympic medal winners, finalists, semifinalists, and qualifying heat swimmers. **Significant difference (p < .01) between males and females. Values are mean ± SE.

Reprinted, by permission, from I. Mujika, S. Padilla, and D. Pyne, 2002, "Swimming performance changes during the final 3 weeks of training leading to the Sydney 2000 Olympic Games," *International Journal of Sports Medicine* 23: 585.

sum of its parts, specific event performance is the most suitable measure to evaluate the effectiveness of a tapering intervention. This is the reason why these authors excluded from their meta-analysis of the effects of tapering on performance all studies using measures other than actual competition or field-based criterion performance data to assess performance capacity. When Bosquet and colleagues interpreted the effect size of taper-induced changes according to the scale of Cohen (1988), expected performance improvements were most often small and occasionally were moderate. When expressed as a percentage difference, mean performance improvement was 1.96%. This difference could be considered meaningless if the population of interest was not competitive athletes. However, the gains that can be expected following a taper intervention, as little as they are, may have a major impact on an athlete's success in major competitions (Bosquet et al. 2007), as illustrated by the preceding example of swimming events at the Sydney 2000 Olympic Games (Mujika et al. 2002b).

AT A GLANCE   **Why Taper?**

Well-designed tapering strategies usually improve performance, but they do not work miracles! A realistic performance goal for the final taper should be a competition performance improvement of about 3% (usual range 0.5-6.0). A performance gain of this size, as small as it may seem, can have a major impact on competition placing no matter what the sport, the event distance, or the caliber of the athlete.

Chapter Summary

When designing a tapering program, coaches and athletes should be aware of the athlete's performance level so that realistic performance targets can be set and the effectiveness of the taper can be assessed after the major competition. Setting unrealistic expectations may have a negative impact on the athlete's psychological state and motivation, so these should always be avoided.

Times achieved in key training sets may be an excellent index of an athlete's adaptation to the taper, and constant communication between athlete and coach can provide more valuable information about how things are going than any biological marker or psychological scale. Based on these simple tools, the taper program can be individualized and refined.

Coaches and athletes should never underestimate the consequences of tapering-induced performance gains, no matter how small these may seem: A gold medal can be just a few centimeters or fractions of a second away!

chapter 7

Insights From Mathematical Modeling

The most important goal for coaches and athletes is to increase the physical, technical, and psychological abilities of the athletes to the highest possible levels of performance and to develop a precisely controlled training program to ensure that the maximal performance is attained at the right moment of the season (i.e., at each point of a major competition). But how do we assess an athlete's adaptation to training throughout the training process?

Applying Systems Theory to Tapering Research

In an attempt to go beyond descriptive experimental procedures to analyze the results of training, Banister and colleagues conducted mathematical analyses of the relationship between performance and the training program (Banister et al. 1975, Calvert et al. 1976, Morton et al. 1990). This type of model analysis is based on systems theory. As indicated by Busso and Thomas (2006), systems theory intends to abstract a dynamic process into a mathematical model. The system is an entity characterized by at least one input and one output, which are related by a mathematical law called transfer function (figure 7.1). The output corresponds to the system's response to the stimuli represented by the input, and the transfer function characterizes the behavior of the system on the basis of parameters estimated from real-life observations.

When applied to athletic performance, this type of mathematical modeling intends to describe training responses based on a whole-body approach that considers performance as the system's output, which varies according to

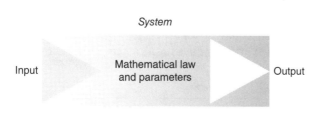

Figure 7.1 System characterized by one output reacting to one input in accordance with a mathematical law, named transfer function, and a set of parameters estimated from observations.

Reprinted, by permission, from T. Busso and L. Thomas, 2006, "Using mathematical modelling in training planning," *International Journal of Sports Physiology and Performance* 1(4): 401.

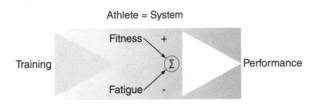

Figure 7.2 A systems model of training. All model variables are measured in the arbitrary unit of the training impulse. The criterion performance is converted to a points score.

Adapted, by permission, from E.W. Banister and J.R. Fitz-Clarke, 1993, "Plasticity of response to equal quantities of endurance training separated by non-training in humans," *Journal of Thermal Biology* 18: 588.

past training, which of course represents the system's input (Busso and Thomas 2006). Various studies have indicated that the development of such mathematical models can improve our understanding of the effects of tapering in optimizing athletic performance (Busso 2003, Fitz-Clarke et al. 1991, Mujika et al. 1996a, Thomas and Busso 2005, Thomas et al. 2008, Thomas et al. in press).

The relevance of a training reduction to maximize performance has been indicated by the modeling of performance variations according to the training load undertaken over time. This approach assumes that performance is the balance between two antagonistic training effects: a positive effect ascribed to fitness adaptations and a negative one ascribed to fatigue (figure 7.2). The model incorporates parameters that are constants, unique to individuals, and it characterizes the response to training of a given athlete. The model parameters are fitted for each athlete from their response to a given training program. The knowledge of these parameters allows the computation of fatigue and adaptation indexes over a specified duration of training (Thomas and Busso 2005, Thomas et al. 2008).

This type of application was used by Mujika and colleagues (1996a) on the training data of competitive swimmers to explain the taper-induced gains in performance (see Aims of the Taper in chapter 1 and Duration of the Taper in chapter 5). This approach was also used to identify the theoretical optimal duration of the taper (Fitz-Clarke et al. 1991) and to establish that a progressive reduction of training can be more effective than a step reduction, a notion in agreement with data observed in triathletes (Banister et al. 1999, Zarkadas et al. 1995) (see Tapering Models in chapter 1). Another possible application is the simulation of changes in performance for variations in training other than that actually done (Fitz-Clarke et al. 1991, Morton et al. 1990, Morton 1991, Thomas and Busso 2005).

AT A GLANCE   **Systems Model**

Mathematical models have been developed to describe the relationship between training and performance. In this approach, the actual training load undertaken by an athlete is the system's input, her actual performance is the system's output, and the positive and negative influences of training (i.e., fitness and accumulated fatigue) are the functions that characterize the system's behavior relating training and performance. The model incorporates parameters that are constant and characterize the specific response to training of individual athletes. Knowing these parameters, investigators can use the model to explain the effects and the optimal duration of a taper and to simulate the effects on performance of a theoretical training load.

Fatigue and Adaptation Model

Taper effects can be described in model studies by the difference in the time course of fatigue and adaptation induced by exercise bouts. Training and performance measurements were undertaken to establish a dose–response relationship based on the negative and positive influences of training (Banister et al. 1975). Model performance was considered to be the balance between these two antagonistic functions ascribed, respectively, to fatigue (negative) and adaptation (positive). A controlled experiment in which total work was reduced after 28 days of strenuous training showed that repeated training bouts could yield a fatigue accumulation greater in magnitude than adaptation (Morton et al. 1990). A greater increase in fatigue than adaptation provoked a transient decrease in performance during the time of intensified training. A subsequent reduction of training loads allowed fatigue (negative influence) to dissipate more quickly than adaptation (positive influence), yielding to criterion performance peaking (Morton et al. 1990). Figure 7.3 illustrates this concept of growth and decline of fatigue and fitness in response to training stimuli.

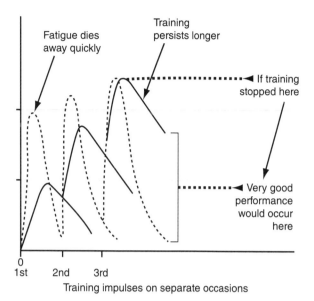

Figure 7.3 Growth and decay of fitness and fatigue in response to impulses of training on separate occasions.

Reprinted, by permission, from E.W. Banister and T.W. Calvert, 1980, "Planning for future performance: Implications for long term training," *Canadian Journal of Applied Sports Science* 5: 172.

As discussed in chapter 1, training responses were also modeled by Mujika and colleagues (1996a) in 18 elite swimmers over a full competitive season including three tapers. This study showed that a progressive reduction of training over 3 or 4 weeks alleviated the fatigue that had accrued with repeated training without compromising the athletes' adaptations. The enhancement of competition performance observed with the taper was mainly attributed to a reduction of the negative influence of training, although small gains in the positive influence of training could also be observed, which undoubtedly contributed to the observed improved performance (figure 7.4).

This general scheme for the effects of tapering should be applicable to a large range of sports, because the model adequacy has also been tested in sports as varied as long-distance running (Banister and Hamilton 1985), triathlon (Banister et al. 1999), weightlifting (Busso et al. 1990, Busso et al. 1992), and hammer throwing (Busso et al. 1994). The goodness-of-fit indicators showed that the model structure allowed an acceptable description of the responses to training across these different sports.

Providing an alternative to Banister's model, Avalos and colleagues (2003) used a linear mixed model to study the training effects on performance in a group of highly trained swimmers over three consecutive seasons. With this modeling technique, instead of constructing a personal model for each athlete, investigators construct a model of popular behavior, allowing parameters to vary from one individual to another, to take into account subject heterogeneity (Avalos et al. 2003). The investigation highlighted the individual nature of short-term, mid-term, and long-term training adaptations as well as the variable nature of an athlete's reaction to a given training stimulus. Individual adaptation to training is not stable over time, so that an identical reaction to an identical training load reiteratively applied through time should not necessarily be expected (Avalos et al. 2003).

In a subsequent investigation, the same authors modeled the residual effects and threshold saturation of training in Olympic-caliber swimmers (Hellard et al. 2005). To determine the residual effect of training, the investigators conducted a multiple regression analysis using performance and several training variables in the short term (weeks 0, 1, and 2 before competition), intermediate term (weeks 3, 4, and 5 before competition), and long term (weeks 6, 7, and 8 before competition). A modified model that includes threshold saturation above which training does not elicit further adaptations was also tested in the investigation. The main outcome of this approach with regard to tapering and peaking was that a low-intensity training load during the taper was related to performance by a parabolic relationship (figure 7.5), indicating that low-intensity training

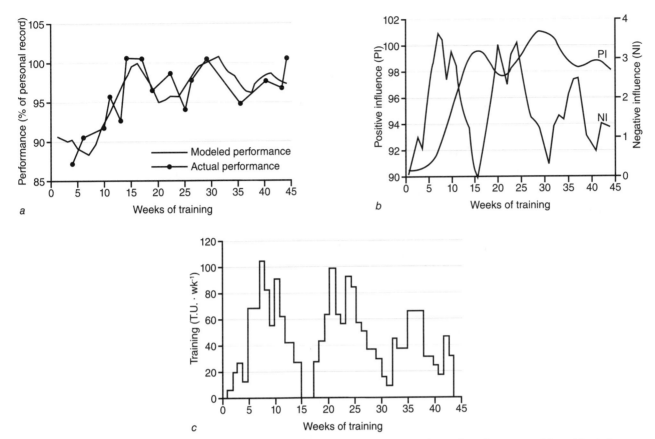

Figure 7.4 Application of the model in one subject: *(a)* fit between modeled and actual performance; *(b)* positive influence (PI) and negative influence (NI) profiles; *(c)* weekly amount of training. PI and NI expressed in the same type of unit as that used for performance. T.U. = training unit.

Reprinted, by permission, from I. Mujika, T. Busso, L. Lacoste, et al., 1996, "Modeled responses to training and taper in competitive swimmers," *Medicine & Science in Sports & Exercise* 28: 255.

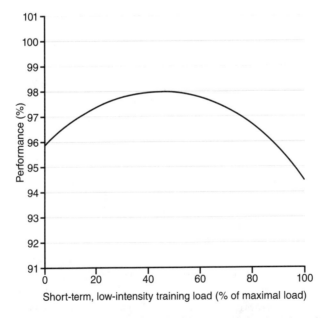

Figure 7.5 Parabolic relationship between short-term, low-intensity training load and performances for the whole group of swimmers. Performance on the vertical axis is expressed as a percentage of the personal record of each athlete. Training load on the horizontal axis is expressed as a percentage of the maximal training load performed by each athlete during the course of the study.

Adapted, by permission, from P. Hellard, M. Avalos, G. Millet, et al., 2005, "Modeling the residual effects and threshold saturation of training: A case study of Olympic swimmers," *Journal of Strength and Conditioning Research* 19(1): 71.

should decrease to 40% to 50% of the maximal training load during the taper phase, whereas lesser and further decreases may jeopardize optimal performance outcomes (Hellard et al. 2005). This is in agreement with observations reported by other investigators in various sports (see chapter 5).

Fatigue and Adaptation Model

AT A GLANCE

According to the fatigue and adaptation model, fatigue increases more than adaptation in response to a training bout. When the training load is reduced, fatigue fades away faster than adaptation, allowing performance peaking. When applied to elite athletes, the model shows that performance peaks during the taper mainly through fatigue dissipation, although adaptation can also increase and contribute to enhanced performance. Mathematical models have also highlighted the individual nature of the adaptation to the taper and the instability of this adaptation over time (i.e., that the same taper program will not necessarily elicit the same adaptation every time).

Variable Dose–Response Model

Banister and colleagues' (1975) model described the individual responses to physical training in terms of adaptation and fatigue, assuming that performance is the balance between these two factors. Those responses depended on model parameters (constants) and were fitted for each participant from his actual response to a training program. Estimation of these parameters permitted the simulation of the responses to a training schedule other than that actually undertaken (Morton 1991). This approach provided useful information on the optimal duration of the taper, typically between 2 and 4 weeks in competitive swimmers (Mujika et al. 1996a), and showed that a progressive reduction of training was more efficient than a step reduction (Banister et al. 1999).

However, the derivations from the model proposed by Banister and colleagues (1975) are questionable because this model assumed that the response to a given training dose was independent of the accumulated fatigue with past training. This also implies that taper durations should be identical whatever the severity of the training preceding the taper. Moreover, according to the original model, any training undertaken around 2 weeks before a competition would be detrimental to performance, so training should be completely stopped for the duration of the taper to maximize performance (Fitz-Clarke et al. 1991). These assumptions are in clear contradiction with evidence-based scientific knowledge on detraining (Mujika and Padilla 2003b) and tapering (Mujika and Padilla 2003a, Mujika et al. 2004). Therefore, the linear formulation of the model with time-invariant parameters seems to provide an imperfect description of training-induced fatigue when a marked change in training regimens occurs (Busso et al. 1997, Busso et al. 2002, Thomas and Busso 2005).

To overcome some of the limitations of the early linear model, a recursive least-squares algorithm was applied to allow the parameters of Banister's model to vary over time (Busso et al. 1997, Busso et al. 2002). These initial studies showed that the responses to a single training session could vary according to the difficulty of the previous sessions. This led Busso (2003) to propose a nonlinear model formulation. The new formulation took into account the accumulation of fatigue with training, on the assumption that the increase in fatigue induced by a given training session was dependent on the severity of prior training sessions. A training session following repeated demanding sessions would induce a greater fatigue than the same session included during training with lower training loads. In other words, the magnitude and duration of the fatigue produced by a given training dose increased with the repetition of training sessions and were reversed when training was reduced (Busso 2003). A downstream implication of this observation is that restoring the athlete's tolerance to training could contribute to the performance peaking induced by a taper.

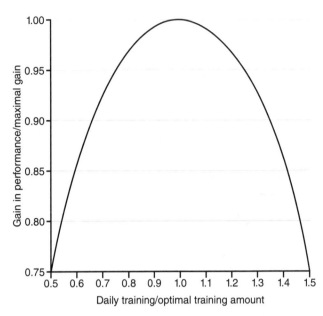

Figure 7.6 Gain in performance at steady state for a same training amount repeated each day. Performance gain was expressed according to its maximal value. Training amount was referred to optimal training yielding maximal gain in performance.

Reprinted, by permission, from T. Busso, 2003, "Variable dose-response relationship between exercise training and performance," *Medicine & Science in Sports & Exercise* 35: 1188-1195.

This nonlinear model was validated using data from six previously untrained subjects enrolled in a controlled training program on a cycle ergometer (Busso 2003). The study led to the conclusion that the nonlinear model described the responses to training more precisely than did previous linear model formulations, also suggesting an inverted-U-shaped relationship between daily amounts of training and performance. This is shown in figure 7.6: When the amount of training exceeds the optimal level, performance gain could decline because of the fatigue induced by oversolicitation. Accumulation of large amounts of intensive exercise with insufficient recovery between training bouts could elicit performance gains smaller than expected. Similarly, a training dose below the optimal amount would elicit smaller performance gains compared with those elicited by an optimal training dose. The nonlinear model is consequently a better approach for the study of the responses to training, in particular the taper after a brief time of overload training (Busso 2003).

The nonlinear model of Busso (2003) has been applied on elite swimmers in real training conditions to describe the athletes' responses to their training and to investigate the influence of the pretaper training on the characteristics (duration, extent, and form of the training reduction) and effectiveness (performance gain) of the optimal theoretical taper (Thomas et al. 2008). The data used were recorded over two entire seasons from four female and four male national and international-standard swimmers who specialized in 100 m and 200 m events. The modeled performances showed a significant fit with the real performances recorded over the two seasons for the eight swimmers. Because the goodness of fit indicates how well a model describes the observed data (figure 7.7) (Busso and Thomas 2006), the adequacy of the fit between modeled and actual performance in this real-life training and competition situation supports the underlying theory of the nonlinear model, which is that a training session well endured during habitual training could be more difficult to cope with when training is intensified, and conversely a more effective training response could be expected when training is reduced and accumulated fatigue fades away, contributing to performance peaking (Busso and Thomas 2006).

The individual model parameters obtained by Thomas and colleagues (2008) allowed analysis of the real training undertaken by the eight swimmers by estimating the variations of the positive and negative influences of training on performance during the two seasons. Figure 7.8 illustrates the results of the application of the mathematical model for one participant. The response to training of this swimmer, which was representative of the whole group, indicates that he achieved his best performances after a taper when the negative influence was reduced and the positive influence elevated (Thomas et al. 2008).

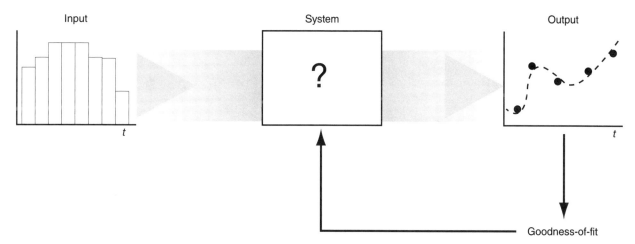

Figure 7.7 Procedure to specify one model for characterizing a dynamic process. Assuming formulation of the transfer function, the model parameters are estimated by fitting model output to actual data. The adequacy of the fit allows the investigator to judge the model's relevance for describing the observation.

Reprinted, by permission, from T. Busso and L. Thomas, 2006, "Using mathematical modelling in training planning," *International Journal of Sports Physiology and Performance* 1(4): 402.

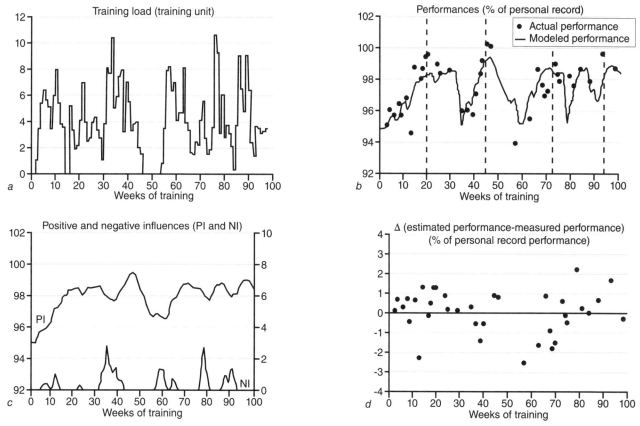

Figure 7.8 Application of the model on swimmer no. 6: *(a)* actual daily amount of training mean averaged on each week; *(b)* fit between modeled and actual performance; *(c)* profiles of positive and negative influences of training on performance (PI and NI, respectively). PI and NI are expressed in the same unit as that used for performance. T.U. = training unit. The vertical dotted lines show the levels of training PI and NI when the swimmer achieved his best performances; *(d)* the residual difference between actual and estimated performance over time.

Reprinted, by permission, from L. Thomas, I. Mujika, and T. Busso, 2008, "A model·study of optimal training reduction during pre-event taper in elite swimmers," *Journal of Sports Sciences* 26(6): 649. Permission conveyed through Copyright Clearance Center, Inc.

The model, however, appeared to underestimate some of the better individual performances and overestimate the poorer performances. The goodness of fit in these highly trained participants in real training conditions was poorer than that obtained with data from less-fit participants in controlled laboratory training (Busso 2003). Possible reasons for these discrepancies are probably a more frequent evaluation of performance under standardized laboratory conditions in the nonathletes and the complex nature of the real-life training undertaken by elite athletes, which precludes an accurate quantification of all training variables included in their training programs (Hellard et al. 2006, Mujika et al. 1996a, Taha and Thomas 2003, Thomas et al. 2008).

These limitations suggest that despite the insights into basic training and adaptation knowledge that can be gained from modeling methodology, it cannot be considered a fundamental tool for coaches and athletes to use when designing training programs. As recently indicated by Busso and Thomas (2006), the strong simplifications used when attempting to model a dynamic process such as the physiological adaptations to athletic training would be barely compatible with the requirements of designing a training program.

As suggested thus far, the variable dose–response model formulation has provided an alternative explanation of the effects of the taper. This model explains performance peaking by accounting for the positive influence of training done during the taper (Mujika et al. 2004, Thomas and Busso 2005). Model computations show that transient decreases in performance with intensified training are attributable to a change in the responses to the training dose (Busso 2003), whereas a progressive reduction of training loads allows an athlete to respond more effectively to training. Therefore, performance peaking with the taper arises from both the recovery from past training and restoration of the tolerance to training (see also Aims of the Taper in chapter 1). The intensification of training delays the positive influence of training. A subsequent reduction of training enhances performance because of the combination of delayed responses to past training and early reaction to training done during the taper. This observation highlights the importance of maintaining sufficient amounts of training during the taper.

From a physiological point of view, the change in dose response between exercise and performance could be related to neuroendocrine responses to training (Mujika et al. 2004). Modulations of the endocrine system and the autonomic nervous system with heavy training could decrease the body's adaptive capacity to training bouts (Kuipers 1998, Viru and Viru 2001). Changes in the neuroendocrine environment with reduced training could modify and amplify the recovery and adaptation process, permitting training loads during the taper phase to maintain or increase the body's adaptation for enhanced performance (Mujika et al. 2004).

AT A GLANCE   Variable Dose–Response Model

The variable dose–response model assumes that the fatigue produced by a given training dose is bigger and lasts longer depending on the severity of prior training sessions and that this effect is reversed when training is reduced. This implies that training tolerance is increased when an athlete is well rested, and thus the training done during the taper can greatly contribute to performance peaking. This model has been shown to better describe the responses to training and tapering than the fatigue and adaptation model and has helped explain performance peaking during the taper as the result of not only recovery from past training but also restoration of the tolerance to training. However, despite the contributions that modeling makes to our understanding of the effects of training programs, modeling methodology is still a long way from being a useful practical tool for coaches and athletes to design specific training and tapering programs.

Computer Simulations

Using previously determined relevant model parameters, investigators can use computer simulations to predict the system's responses to hypothetical solicitations (figure 7.9). It is thus possible to approach a prediction or a control problem by seeking the output for a given input or input for a given output, as long as the system's structure is known (Busso and Thomas 2006).

Computer simulations in modeling the responses to athletic training have been performed in an attempt to gain insight into tapering effects and optimal tapering designs. The parameters obtained in linear modeling studies have been used to simulate the change in performance in response to given variations in training and establish the critical schedule of rest or reduced training before competition (Fitz-Clarke et al. 1991). In addition, the relative effectiveness of step versus exponential tapers and fast versus slow exponential decay tapers in producing optimal improvement was determined by this method, and the results of the simulation were tested experimentally in triathletes (Banister et al. 1999) (see also Tapering Models in chapter 1).

The nonlinear model outputs elicited by Busso (2003) have been used to analyze the factors that influence the optimal taper characteristics, including duration, extent, and form of the training reduction. Thomas and Busso (2005) based their theoretical study on simulations of the responses of the six participants of the study by Busso (2003) to various hypothetical tapering strategies. Thomas and Busso (2005) showed that optimization of the taper requires the best possible compromise between the extent of training reduction and its duration, so that the characteristics of an optimal taper would depend on prior training. According to the simulations, intensified training before the taper facilitated better performances but presumably would demand a greater reduction of the training load over a longer time. Moreover, an optimal progressive training reduction led to a better performance than an optimal step reduction when the taper was preceded by an overload.

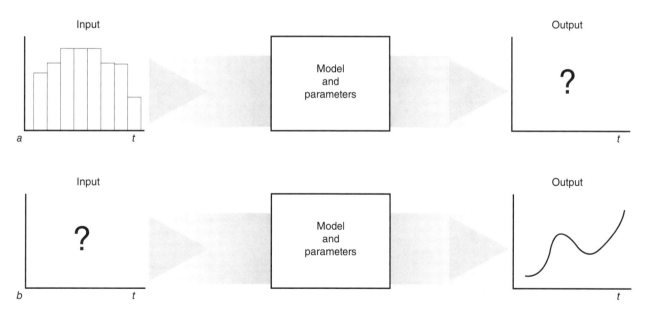

Figure 7.9 Schematic representation of the use of one model to predict the system's behavior under given circumstances. The specification of the model and the estimation of the parameters from previous observation allow two types of theoretical analyses: (a) the assessment of the model output for a given input (prediction) and (b) the research of the input that elicits a given output (control).

Reprinted, by permission, from T. Busso and L. Thomas, 2006, "Using mathematical modelling in training planning," International Journal of Sports Physiology and Performance 1(4): 403.

Nevertheless, the model-derived characteristics of the theoretical optimal taper did not match very well the data arising from the available literature (Mujika and Padilla 2003a). These discrepancies may have arisen from the data used to derive the model, which came from a nonathletic population enrolled in a controlled experiment, in contrast with most studies about the taper, which mainly focus on athletes. Therefore, it has been suggested that the results of that theoretical study deserve to be reexamined by using model parameters obtained in athletes in a real training situation (Busso and Thomas 2006).

That is exactly what Thomas and colleagues did in two additional investigations. Thomas and colleagues (2008) used model parameters arising from eight elite swimmers' real-life training and performance data to undertake computer simulations to reexamine the characteristics of the optimal taper according to prior training. The authors also determined whether the gain in performance with an optimal reduction of training resulted from further adaptations in addition to fatigue dissipation in this elite athletic population. The computer simulations indicated that the optimal taper without prior overload training required a mean training reduction of 65% over 16 days. If the taper was preceded by 4 weeks of intensified training, the extent of the optimal reduction was not significantly different (67%) but had to be maintained over a longer duration (22 days). As previously observed in nonathletic subjects (Thomas and Busso 2005), the highest mean performance reached with the optimal step taper was significantly greater with than without prior overload training (101.4% vs. 101.1% of personal record). The comparison from the beginning of the taper showed that the performance gain during the step taper after overload training (3.60%) was significantly greater than without prior overload (1.99%) (figure 7.10). It was also observed that both with and without overload training before the taper, the two forms of progressive taper (i.e., linear and exponential) required a training reduction significantly smaller, over a longer duration, than the step taper.

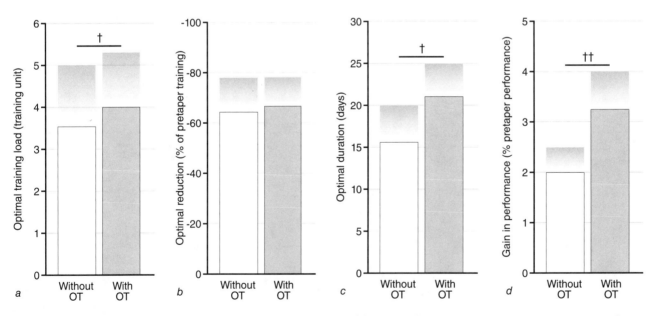

Figure 7.10   Effect of the pretaper training on the characteristics of the optimal step taper. Values are mean ± s_x. Significant difference between taper without and with a prior overload time (OT): †p < .05 and ††p < .01.

Reprinted, by permission, from L. Thomas, I. Mujika, and T. Busso, 2008, "A model study of optimal training reduction during pre-event taper in elite swimmers," *Journal of Sports Sciences* 26(6): 650. Permission conveyed through Copyright Clearance Center, Inc.

The joint analysis of performance and the positive and negative influences of training indicated that although the taper essentially acted through fatigue dissipation, the benefit of a prior overload results from further adaptations occurring during the taper (figure 7.11). As shown in figure 7.11, an optimal taper should completely eliminate the negative influence of training (i.e., accumulated fatigue), even after pretaper overload training. A higher adaptation could be achieved with an overload before the taper, and although pretaper performance level would be hindered by the overload, a bigger performance gain could be achieved after an optimal taper following overload training.

In summary, the computer simulation study on elite swimmers of Thomas and colleagues (2008) showed that an overload before the taper is essential to maximize performance but imposes specific requirements during the taper. A 20% increase in normal training during 28 days before the taper requires a step reduction in training of around 65% during 3 weeks, instead of the 2 weeks required when no overload training is performed. A progressive (linear or exponential) taper is preferable to a step taper after prior overload training, but it should last nearly twice as long as the step taper. These findings confirm the relevance of the original modeling approach in the study of individual responses to training and the optimization of tapering strategies (Thomas et al. 2008).

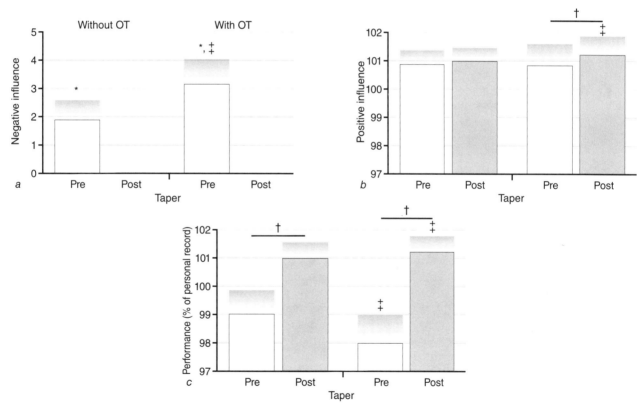

Figure 7.11 Effect of the optimal step taper without and with prior overload training (OT) on *(a)* the negative and *(b)* the positive influences of training and *(c)* on performance. The negative (NI) and positive influences (PI) are expressed in the same unit as that used for performance. Values are mean ± s_x. Statistical difference ($p < .05$): *from 0 (for NI); †between pretaper (Pre) and posttaper (Post), and ‡between values without and with OT.

In a subsequent modeling investigation, Thomas and colleagues (in press) used computer simulations to test the hypothesis that a two-phase taper with a final increase in the training load at the end of the taper is more effective than a traditional linear taper. The main outcome of the investigation was that the last 3 days of the taper were optimized with a 20% to 30% increase in training load. Compared with a traditional linear taper, the performance benefits of such a two-phase taper were very small and were explained by additional adaptations, whereas the removal of fatigue was not compromised (see also Tapering Models in chapter 1).

It is obvious that the theoretical results reported here must be tested experimentally before conclusive training recommendations can be provided to coaches and athletes. Nevertheless, insights arising from mathematical models and computer simulations can have practical implications for athletes and coaches in the lead-up to competitions, particularly multiday athletic events. However, it is also clear that mathematical modeling is a fair way away from becoming a tool to help coaches design and prescribe training programs for particular athletes. To achieve this goal, new modeling strategies need to be developed that overcome the limitations of available models (Busso and Thomas 2006, Hellard et al. 2006, Taha and Thomas 2003).

AT A GLANCE   Using Computer Simulations

Using previously determined model parameters, investigators have used computer simulations to gain insight into tapering effects and optimal tapering designs. Simulations based on linear model parameters have contributed to establishing the optimal taper duration and the suitability of progressive versus step tapers. Variable dose–response model parameters have been used recently to assess optimal taper characteristics of elite swimmers, predicting that increasing the training load by 20% for 4 weeks to overreach the athletes prior to the taper could optimize performance but would require a longer taper to completely dissipate fatigue and elicit further adaptations. Computer simulations also predict that a 20% to 30% increase in the training load during the final 3 days of the taper would not compromise fatigue removal and could improve performance by eliciting additional adaptations. These predictions, of course, require experimental confirmation before conclusive recommendations can be provided to coaches and athletes.

Chapter Summary

Mathematical models have significantly contributed to the understanding and optimization of tapering programs. These models consider that the athlete is a system responding to the training input with fatigue and adaptation and that the difference between these two opposite functions elicits the performance output. The models also incorporate parameters that are different for each athlete. The linear fatigue and adaptation model has been used to determine that performance during the taper peaks mainly through fatigue dissipation and to establish a framework for optimal taper duration. However, linear models consider that a given training load induces the same fatigue response at any time of the season, which is a major limitation. Alternative mathematical models have been applied that highlight the individuality of the adaptation to the taper and the fact that this adaptation changes over time.

Pros and Cons of Mathematical Modeling

Despite the contributions of mathematical modeling to the understanding of athletes' responses to training in general and tapering in particular, modeling methodology needs to be refined before it can become a readily available tool for coaches and athletes to help them design optimal training programs and predict their performance outcomes. Nonetheless, athletes, coaches, and sport scientists should keep in mind Dawkins' (2006) comment: "Of course there are good models of the world and bad ones, and even the good ones are only approximations. No amount of simulation can predict exactly what will happen in reality, but a good simulation is enormously preferable to blind trial and error" (p. 58). In this respect, it is worth mentioning that my colleagues and I used a mathematical model approach in the late 1990s to help coaches design the tapering programs of a group of French elite swimmers, with excellent results. Most of the swimmers attained their best performance of the year at the desired moment of the season, and this approach provided the coaches and athletes with a practical tool that they told me was "enormously preferable to blind trial and error."

A nonlinear variable dose–response model has been developed to overcome the main limitation of the linear fatigue and adaptation model, allowing model parameters to change over time in response to prior training loads. With this approach, the model assumes that a given training load will elicit a different level of fatigue depending on the training previously undertaken by an athlete. This model has contributed to explaining that tapering-induced performance gains are attributable not only to fatigue dissipation but also to further fitness gains that are allowed by an increased tolerance and adaptation to the training performed during the taper, when an athlete is better rested as fatigue fades away.

Another application of mathematical modeling allows us to perform computer simulations to either predict the response to a theoretical training program or determine the training input necessary to achieve a theoretical performance output. These simulations have also contributed to establishing optimal taper durations and models and to making theoretical predictions about how overload training before the taper (i.e., overreaching the athletes) or final increases in the training load at the end of the taper would affect tapering program design and performance outcomes. These theoretical predictions need to be confirmed or rejected experimentally.

chapter 8

Unique Aspects of Team Sport Tapering

The periodization of training is a method that allows athletes from individual sports to peak at the most important competitions of the season. These athletes usually achieve a fitness and performance peak through months of hard work followed by a segment of tapered training, culminating with the targeted race or championship. As we have seen in previous chapters, many of the physiological, psychological, and performance benefits of such a strategy are now well established. However, this approach may not always be the most suitable for team sport athletes. Indeed, whereas individual sport athletes can afford to perform below expectations and even miss competitions that do not fall within the scope of their major goals, although always in the best interest of such major goals, team sport athletes usually need to perform at a high level week after week if they want to be in contention for the championship when it really counts.

Why So Little Research on Tapering and Peaking in Team Sport?

As recently reviewed by Pyne et al. (2009), most of the experimental and observational research on tapering in the scientific literature has been conducted primarily in individual (predominantly endurance) sports and events. There are two main reasons for the focus on individual sports: First, in these sports there are moderate to large correlations between physiological capacities, basic training factors such as volume and intensity, and competitive performance; second, these factors are much easier to isolate and quantify in comparison with the multifaceted nature of team sport's physiological demands, training, and performance (Mujika 2007a). Team sports clearly involve a combination of physical, physiological, psychological, technical, and tactical factors that contribute to performance. Given that most team sports require well-developed speed, acceleration, power, endurance, and agility, it seems likely that an effective taper would improve many or all of these attributes (Mujika 2007b).

In short, it seems that the worldwide popularity of team sports among fans and the general public does not have a parallel among sport scientists. The contributions from sport science to the understanding and optimization of tapering and peaking strategies directly related to team sports are indeed scant.

In a 2007 editorial, Mujika (2007a) suggested that a possible reason for this apparent paucity of team sport research is that this type of research is distinctly difficult to carry out. In that piece, Mujika listed several major problems that anyone seeking to do research on tapering and team sports must overcome:

- The physiological determinants of team sport performance are not clearly understood in comparison to most individual sports involving different types of locomotion, such as running, swimming, cycling, or rowing, or various modes of jumping, throwing, or lifting. Identifying physiological qualities is a *sine qua non* but clearly not the only requirement for athletes to be competitive in the team sports arena.

- Performance itself is a difficult concept to define in the world of team sports. What is performance in water polo, football, or basketball? Scoring more goals or points? Maintaining a higher playing tempo than the opposition for the duration of a match? Being able to demonstrate skills and qualities under the high pressures of actual competition? Sport scientists are used to dealing with precise, quantifiable, numerical data, and although these can be indicators of an athlete's potential to perform, actual performance within a team sport framework is a relatively abstract concept.

- Quantification of training is a key aspect of high-quality sport physiology research, particularly to assess the influence of training loads on physiological responses and adaptations and the relationships between these measures and performance capabilities. But team sport training generally involves a diverse range of training activities, often under quite variable environmental conditions, as well as interindividual variability in the responses and adaptations to training. These issues complicate the integration of training variables into quantifiable units.

- Another potential difficulty for team sport research is the long competitive seasons within and between national and international competitions. For instance, a player from any of the major European football clubs usually competes domestically (league and cup) and internationally (Champions League or UEFA cup) from mid-August to mid-May or June. Every other year domestic competition is immediately followed by national team competition, such as European or World Cup. Given this scheduling, any given elite level player takes part in more than 60 matches during the season. Under these conditions, it is not easy to carry out experimental research that would place additional physical demands on the already overloaded players.

- The relatively high risk of injury associated with many team sports makes it difficult to carry out longitudinal investigations during the competition season.

Despite the great difficulties inherent in team sport research, sport scientists working closely with team sports should begin to investigate possible solutions to these problems. Only then will they be able to make much-needed contributions to this somewhat forgotten but extremely exciting area of sport physiology, tapering, and performance peaking. This chapter discusses the limited research that has been done despite the difficulties just described.

Single and Multiple Peaking

No study has directly examined the taper in the context of multiple peaking. In most individual sports, the major competition involves a series of heats, semifinals, and finals that can stretch over several days, even several weeks, as in the grand tours of professional road cycling. In team sports, most national and regional competitions involve one or more games per week over a 4- to 8-month season. Because of the lack of research in the area of multiple peaking, it is not known how often an athlete or team can obtain the performance benefits of an efficient taper, and it is therefore not possible to make sound recommendations in this respect (Pyne et al. 2009). Another key consideration in team sports is preparation of national teams for major international competitions like the World Championships, World Cup, or Olympic Games. Team-sport tapers include variants for regular season peaking and major tournaments (Mujika 2007b). The physical preparation of players that is prescribed in these settings has evolved through trial and error rather than experimental research.

The scarcity of published studies highlights the potential for team sports to gain a competitive advantage if they can develop effective strategies for tapering prior to competition. Two competitive situations come to mind with regard to tapering and peaking for team sports: preseason training to face a league-format competitive season in the best possible condition, and peaking for a major international tournament such as the Olympic Games or World Championships.

Tapering and Peaking for the Regular Season

Most coaches understand that having all players in a team at peak fitness levels for the duration of a season is simply not possible. But it is important to prescribe periodized conditioning programs in the preseason to make sure that players' physical capacities are optimal at the onset of the competitive season.

In a recent investigation, Coutts and colleagues (2007a) examined the influence of preseason deliberate overreaching and tapering on muscle strength, power, endurance, and selected biochemical responses in semiprofessional rugby league players. The athletes completed 6 weeks of progressive overload training with limited recovery, followed by a 7-day progressive taper. Following the overload segment, multistage fitness test running performance was significantly reduced (12.3%), and most other strength, power, and speed performance measures tended to decrease (range –13.8% to –3.7%). Significant changes were also observed in selected biochemical markers such as plasma testosterone to cortisol ratio, creatine kinase, glutamate, and glutamine to glutamate ratio ($p < .05$). Following the taper, a significant increase in peak hamstring torque and isokinetic work was observed as well as minimum clinically important increases in the multistage fitness test, vertical jump, 3RM squat, 3RM bench press, chin-up, and 10 m sprint performance. Moreover, all biochemical markers tended to return to baseline values. The authors concluded that muscular strength, power, and endurance were reduced following the overload training, indicating a state of overreaching. The most likely explanation for the decreased performance was increased muscle damage via a decrease in the anabolic–catabolic balance. However, it was shown that a subsequent progressive taper may induce supercompensation in muscular strength, power, and endurance that is related to increased anabolism and a decrease in muscle damage (Coutts et al. 2007a).

Repeated sprint ability, which is a basic performance requirement for most team sports, can also be enhanced through periodized training and tapering. Bishop and Edge (2005) investigated the effects of a 10-day exponential taper subsequent to 6 weeks of intensive training on repeated-sprint performance in recreational-level, team-sport female athletes. Subjects were tested for repeated sprint ability (5 × 6 s all-out cycling sprints every 30 s) before and after the tapering segment. As shown in figure 8.1, the 10-day taper resulted in a nonsignificant decrease in total work (4.4%; $p = .16$) and peak power (3.2%; $p = .18$) and a significant decrease in work decrement (10.2% ± 3.5% vs. 7.9% ± 4.3%; $p < .05$).

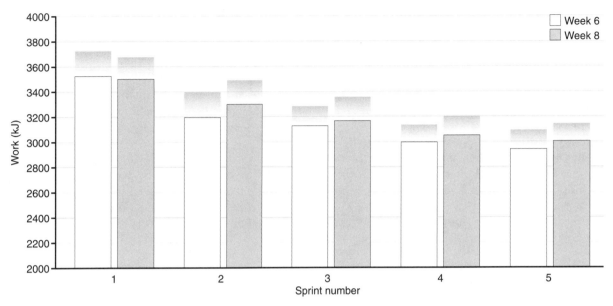

Figure 8.1 Total work (kJ) recorded for each sprint of the 5 × 6 s sprint test before (week 6) and after the 10-day taper procedure (week 8). Data are mean ± SEM.

Reprinted, by permission, from D. Bishop and J. Edge, 2005, "The effects of a 10-day taper on repeated-sprint performance in females," *Journal of Science and Medicine in Sport* 8(2): 205.

AT A GLANCE   **A Team Is Also a Group of Individual Athletes**

The investigations described in this chapter show that basic training principles that can be effective in individual sport athletes also apply in team sports. Coaching and conditioning staff should therefore take these principles into account when designing their preseason training programs.

Once the season starts, how a team maintains the peak fitness levels achieved by a successful preseason periodized training program will depend on several factors, such as time between games, travel, competitiveness of the opposition, injury, minutes of match play, physiological adaptations to competition, recovery, and training of individual players. All these variables must be integrated into the in-season training plan for the team to retain or further improve early-season fitness and performance levels.

Tapering and Peaking for a Major Tournament

It seems that professional football (or soccer in the United States) players competing for their clubs in the lead-up to major international tournaments such as the World Cup, and therefore having reduced opportunities to taper, are among those most likely to underperform (Ekstrand et al. 2004). Most major international tournaments take place at the end of a long club-level competitive season. In an attempt to elicit players' peak performance, some nations decide to advance their domestic competition calendar to allow the players to rest and rebuild their fitness to compete for their national teams. A different approach is to delay the end of the domestic season so that the players are still in a competitive shape when they join their national squad. Both strategies have pros and cons, and the scarce scientific literature available is not conclusive regarding the optimal approach to peaking for a major tournament.

Bangsbo and colleagues (2006) recently described the preparation program of the Danish national football team for the 2004 European Championship. After the club season, the players rested for 1 to 2 weeks before preparing for the championship. The preparation lasted 18 days divided into two 9-day phases (table 8.1).

Table 8.1 Training Schedule for Two 9-Day Periods (Phases 1 and 2) for the Danish National Soccer Team Before EURO 2004

Day	PHASE 1 Morning	PHASE 1 Afternoon	PHASE 2 Morning	PHASE 2 Afternoon
1	Yo-Yo IE2 test Technical and tactical training	Aerobic$_{HI}$ training (6 × 2 min) Play—20 min	Yo-Yo IE2 test Technical and tactical training	Speed training Technical and tactical training Play—20 min
2	Free	Technical and tactical training	Free	Aerobic$_{HI}$ training (6 × 2 min) Technical and tactical training Play—20 min
3	Technical and tactical training	Speed training Technical and tactical training Speed endurance maintenance training	Technical and tactical training	Speed training Technical and tactical training
4	Free	Technical and tactical training Play—30 min	Group C: Speed endurance Production training	Friendly game (evening)
5	Free	Speed training Aerobic$_{HI}$ training (8 × 2 min)	Free (traveling)	Free (traveling)
6	Free	Technical and tactical training Group C: Aerobic$_{HI}$ training (6 × 2 min)	Aerobic$_{HI}$ (3 × 5 min) Play—30 min	Free
7	Free	Friendly game	Technical and tactical training	Speed training Technical and tactical training Speed endurance production training
8	Free	Group A: Recovery training Group B: Speed training Play—30 min	Free	Technical and tactical training
9	Free	Aerobic$_{HI}$ training (8 × 2 min) Play—20 min	Yo-Yo IE2 test Technical and tactical training	Speed training Technical and tactical training Play—20 min

Reprinted, by permission, from J. Bangsbo, M. Mohr, A. Poulsen, J. Perez-Gomez, and P. Krustrup, 2006, "Training and testing the elite athlete," *Journal of Exercise Science and Fitness* 4(1): 1-14.

The amount of high-intensity exercise was similar in both phases (i.e., training intensity was maintained), whereas the total amount of training was reduced in the second phase (i.e., training volume was tapered) (figure 8.2). This is in agreement with previous tapering recommendations based on studies from individual sport athletes (Mujika and Padilla 2003a).

The authors emphasized that because of large individual differences among players in the amount of high-intensity work performed during the tactical components of the training sessions, a careful evaluation of individual physical training load is essential, even during training time not specifically dedicated to fitness development (figure 8.3).

Ferret and Cotte (2003) reported on the differences in preparation of the French national football team in the lead-up to the World Cups of 1998 and 2002. The former World Cup campaign saw Les Bleus taking home the valued trophy. Four years later, an almost identical group of players returned home sooner than expected, after a disappointing qualifying round without a single victory and not scoring a single goal. According to these authors, in 1998 the team had enough time and biological resources prior to the qualifying round to further develop the athletic qualities of the players through

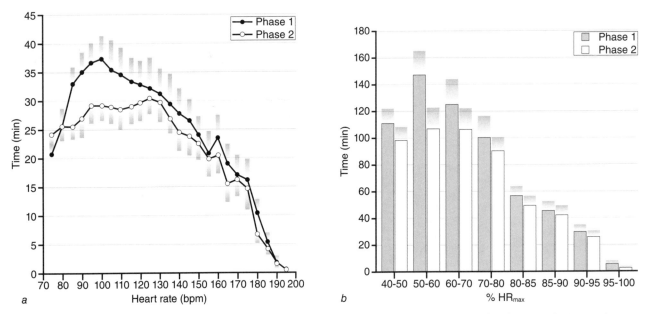

Figure 8.2 Heart rate distribution during two 9-day preparation phases (phases 1 and 2) for the Danish National team soccer squad before the European Championship 2004. The values are expressed as mean ± SEM in *(a)* beats per minute and *(b)* percentage of maximum heart rate

Reprinted, by permission, from J. Bangsbo, M. Mohr, A. Poulsen, J. Perez-Gomez, and P. Krustrup, 2006, "Training and testing the elite athlete," *Journal of Exercise Science and Fitness* 4(1): 1-14.

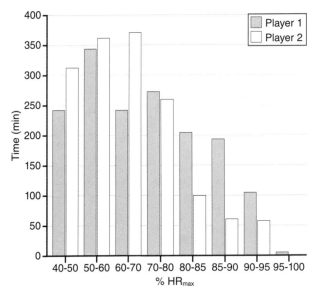

Figure 8.3 Heart rate distribution of two players in the Danish national team soccer squad during an 18-day training segment (sum of phases 1 and 2) before the European Championship 2004.

Reprinted, by permission, from J. Bangsbo, M. Mohr, A. Poulsen, J. Perez-Gomez, and P. Krustrup, 2006, "Training and testing the elite athlete," *Journal of Exercise Science and Fitness* 4(1): 1-14.

two solid training phases followed by a 2-week tapering phase, characterized by high-intensity training situations (friendly games) and a moderate training volume, which allowed players to eliminate the negative effects of training (fatigue) while maintaining the adaptations previously achieved. In contrast, in 2002 all players were only available to the national team 8 days prior to the beginning of competition, and medical and biochemical markers indicated that most players were severely fatigued after the club season. In those conditions, the technical staff could not carry out a development training phase followed by a taper to peak the physical qualities of the players prior to the World Cup (Ferret and Cotte 2003).

The reports just described suggest that an ideal approach to peak for a major international tournament would start several weeks before the first game, with an initial recovery after the club season, followed by rebuilding, and finishing with a pretournament taper characterized by low training volume and high-intensity activities.

Nevertheless, there are examples of successful unorthodox approaches that challenge these ideas about optimal preparation. For instance, the Danish national football team unexpectedly won the 1992 European Championship after the team was invited to compete 10 days before the beginning of the tournament, because of the last-minute exclusion of Yugoslavia from the championship. By then, half the Danish players had already finalized their participation in various European leagues and had been out of training for 3 to 5 weeks, whereas the other half were still competing in the Danish domestic championship. All players were only available to the coaching staff 6 days before the first game. The team's success has been partly attributed to the fact that players were not physically and psychologically exhausted, as is often the case after long and tough domestic and international club seasons followed by a long national team preparation and a demanding international tournament (Bangsbo 1999).

Major Tournament Preparation for Teams ⬛ AT A GLANCE

Although preparing teams for major tournaments is not, in principle, radically different from preparing athletes in individual sports for championship events, the complexities of a group taper can be daunting. Still, the basic structure of a segment of relative rest following the regular season followed by a time of serious training culminating in a taper is a reasonable strategy for team coaches to adopt. As with individuals, the team taper should feature no reduction in intensity but should entail a low training volume. Exceptions to this well-ordered plan exist, as the experience of the Danish European Football Championship team of 1992 illustrates. Creativity and making the most of the unexpected are hallmarks of the most successful athletes and their coaches.

Chapter Summary

Clearly there are different approaches to optimizing team sport performance in the lead-up to a major tournament. The coaching staff should consider biological, technical, tactical, psychological, and sociological variables when deciding on the most suitable strategies to get the best performance from their players.

Team sport athletes generally need to perform at a high level for long periods of time. Tapering and peaking strategies for team sports are not as well researched and developed as they are in individual sports, but most established tapering principles for individual athletes also apply to team sport athletes.

Intensive preseason training programs should finish with a taper to facilitate supercompensation and ensure close-to-optimal values in fitness traits that are relevant to team sport performance, such as strength, power, endurance, and repeated sprint ability.

Peaking for major international tournaments usually poses the problem of choosing between recovering from domestic competition and then rebuilding players' fitness, or maintaining intensive training and capitalizing on adaptations acquired during the domestic season. Both approaches can be valid, and the choice should depend on the level of fatigue the players present after the domestic season and the amount of time between the end of domestic competition and the beginning of national team competition.

Additional information on tapering and peaking in a team sport context is available in chapter 12, in which extremely successful team sport coaches share their views and specific training and performance peaking strategies in the lead-up to major achievements in international competition.

part III

Elite Sports Figures on Tapering and Peaking

Darío Rodriguez

AINHOA MURUA winning the 2007 Spanish Triathlon Championships in Úbeda, Spain. Murua was not favored to win because she had been forced to stop training for several months that year attributable to mononucleosis. Yet she placed a very solid first, being the second fastest swimmer, second strongest biker, and the third fastest runner in the race. Murua is coached by Iñigo Mujika.

In part I we have discussed the theoretical bases of what is known about tapering and peaking for major competitions: what tapering is; the different tapering models that coaches, athletes, and researchers use in their attempt to optimize athletic performance; and the physiological and psychological changes that tapering brings about. In part II, we have devoted four chapters to discussing the training modifications that can be performed during tapering periods, the performance implications of the taper, what we have learned about the taper from mathematical models, and what we know about tapering and peaking for team sports.

It is now time to discuss the more applied aspects of tapering and peaking for competition, and what better way to do this than to hear about it directly from successful coaches and athletes? In part III we hear from some world-class coaches and athletes about their approach to tapering to peak for major competitions. Their contributions have been distributed in four chapters dealing respectively with endurance sports, sprint and power events, precision sports, and team sports.

In contrast with part I and part II, part III does not rest on the solid grounds of scientific research. Therefore, the exact effects of the tapering and peaking strategies on physiological, psychological, and performance variables are not precisely described and quantified. On the other hand, the experiences that these elite athletes and coaches share in the next four chapters reflect the real world of sports and show how various approaches to preparing for competition can be successful on the world stage. Athletes and coaches will identify with some of the views expressed by these very successful peers, will find many useful tips and much food for thought, and will be inspired by the rich resources provided by these world-class sport figures.

chapter 9

Tapering for Individual Endurance Sports

As we can gather from previous chapters, most of the knowledge on tapering available in the scientific literature, whether of an experimental or observational nature, pertains to individual endurance sports and events. These involve different modes of human locomotion, including swimming, cycling, running, rowing, and combined sports such as triathlon. There seem to be two main reasons for this research focus on individual sports: First, in these sports there are moderate to large correlations between physiological capacities, basic training factors such as volume and intensity, and competitive performance. Second, training and performance factors related to individual sports are much easier to isolate and quantify than are the same factors in regard to team sport (Pyne et al. 2009).

Bob Bowman

Tapering World Champion Swimmers

In the competitive world of high-performance athletics, it is paramount that athletes be at their best when their best is needed. In Olympic sports, this means that a peak of physical, psychological, and emotional energy must be attained for the culminating event in each quadrennium, the Olympic Games. With event specialization and the increased use of science and technology to improve athletic performance, it is imperative that the coach and athlete plan for success in both the long and near terms. The following sections explore the steps leading to peaking for major competitions in swimming and how one coach organizes his taper plans.

Planning

One thing becomes clear when we think about achieving performances on the world level: Failing to plan is planning to fail. A coach must take an inventory of all the elements that comprise his quadrennial plan and then shape shorter-term objectives within this broad framework. First and foremost is the athlete. Is he experienced in international competition or a relative newcomer? What are her technical strengths and weaknesses? How about physical strength, power, and endurance? Overall health and ability to withstand heavy workloads must also be taken into consideration. Is the athlete strong mentally, or does he require more seasoning under high-pressure conditions?

BOB BOWMAN

Multiple gold medal-ist Michael Phelps walks with his coach Bob Bowman during a training session for the 2007 World Swimming Championships in Melbourne, Australia.

© AP Photo/Mark Baker

Bob Bowman has coached elite swimmers for more than 20 years. Throughout his career, he has coached a number of swimmers to the Olympic Games and the World Championships, winning gold, silver, and bronze medals at these major events. Michael Phelps (with a total of 14 Olympic Gold Medals to his credit), Peter Vanderkaay (winner of two Olympic Gold Medals and two World Championships Gold Medals), and Erik Vendt (winner of Olympic Gold in 2008 and Olympic Silver in 2000 and 2004) are among the most successful athletes coached by Bob. Among other awards and distinctions, Bob was honored with the USA Swimming and the American Swimming Coaches Association coach of the year awards in 2001 and 2003. He was also named the USA Swimming developmental coach of the year in 2002. Bowman was named Team USA's head coach for the 2007 World Championships in Melbourne, Australia.

Each coach will have a different set of criteria and will plan her quadrennium in different ways to meet the special needs of the athlete. I have always believed that planning must progress from general (long term) to specific (short term) and from simple (macrocycle) to complicated (microcycle):

- **Macrocycles.** After a broad, general goal has been developed for each athlete (i.e., win gold in the 2004 Olympic Games), more specific and detailed programs can be put into place. Each year of the 4-year term will have a major international meet and perhaps another national competition of importance. Thus each training year will be divided into two macrocycles: one culminating in a national championship (or trials meet) and the other in an international competition (World Championships). Within each macrocycle there will be periods of transition, building, intensity, competition, and specific training. Each macrocycle will also include a taper or peaking period, with the international competition macrocycle having a more extensive fine-tuning stage. For an example, see table 9.1, which is Michael Phelps' 2000-2001 annual plan. Once the coach has determined his competition schedule for each macrocycle and has defined the training periods, then he can focus on creating a microcycle plan.

- **Microcycles.** In our program the microcycle plays a critical role because the general outline of training remains consistent within each training week. The emphasis placed on certain qualities (i.e., aerobic endurance, muscle endurance, anaerobic power, and mixed training) determines the theme of training within given periods.

Although the emphasis and volume of the different types of work change as the season progresses, each type of training is included in each microcycle. This provides the foundation of the taper plan and ensures that athletes hit each of the important energy systems within each training block. A standard midseason microcycle for middle-distance swimmers at Club Wolverine would look something like the microcyle shown in table 9.2.

Components of Tapering

The main challenge when designing a peaking plan for an athlete is maintaining control of the various elements that interact with each other during the taper process. For instance, it is unwise to reduce both volume and intensity of work simultaneously. Therefore in our taper program, we first reduce the volume of the work being done while maintaining the same levels of intensity.

This process formally begins approximately 21 days before the start of the first event at the major competition. I take one extra step off approximately 6 weeks prior to the competition.

One way to manage the volume of work is to reduce the number of sessions per week (table 9.3). I often have swimmers go from four to three morning practices in the microcycle about 6 weeks prior to the peak competition. This automatic reduction ensures that they consistently reduce volume particularly in the early stages of the peaking period when the demand for intensity (speed) in the workouts is high. It also provides a more uniform pattern of practices throughout the taper period.

Intensity is the critical element and must be maintained throughout the bulk of the taper period. Only in the final days before the competition, when the volume of work has been reduced to its lowest level, should the intensity of the sessions be reduced to maintenance levels.

Let's take a look at an actual taper plan for an elite athlete (table 9.4). Michael Phelps' final preparations for the 2007 World Championships are presented here. Michael won seven gold medals and set five world records in what has come to be known as the greatest individual performance in the history of our sport, only bettered by Michael's incredible performance a year later at the 2008 Beijing Olympics, where he added an additional gold medal and another world record to his 2007 World Championships achievements.

The plan calls for a gradual reduction in training volume (expressed in 1,000 m increments). Michael's average volume during the heart of the season would be 6,000 m in the morning and 7,000 m in the afternoon. The formal taper program begins at 20 days out and follows a 2-week period where the volume was gradually reduced by about 20% over peak levels. However, the intensity of the work remains constant. The training paces for both threshold and quality work are the same with the only real difference being the number of repetitions or time spent doing each type of training. The ratio of a certain type of work to the total training volume remains unchanged.

The taper plan accounts for a very taxing plane trip to Australia and the resulting adaptation to the time change. Michael is an experienced traveler and takes all necessary precautions to minimize any complications from the change to a different time zone.

Event-Specific Considerations

The coach must understand the individual needs of each athlete in his group when preparing for a major competition. Men and women will have different plans, particularly for dry land training. The men will require much more rest from the intense land work whereas the women must work to maintain their strength throughout the taper. Volumes will be higher for long-distance swimmers than for middle-distance or sprint athletes. Older swimmers will generally require more rest and less volume than younger athletes. In our preparation for the World Championships, Michael Phelps, Peter Vanderkaay, and Klete Keller all followed the same general plan. All swam on the U.S. world record–setting 4 × 200 relay, but each swimmer's plan deviated slightly from the general plan.

Michael spent his time divided between the four strokes because he is primarily an individual medley swimmer. Peter and Klete trained exclusively for the 400 and 200

Table 9.1 2000-2001 Seasonal Plan for Michael Phelps (NBAC)

Dates

Months	Microcycle	Week Begins
September	1	25 September
October	2	2 October
October	3	9 October
October	4	16 October
October	5	23 October
October	6	30 October
November	7	6 November
November	8	13 November
November	9	20 November
November	10	27 November
December	11	4 December
December	12	11 December
December	13	18 December
December	14	25 December
December	15	1 January
January	16	8 January
January	17	15 January
January	18	22 January
January	19	29 January
February	20	5 February
February	21	12 February
February	22	19 February
February	23	26 February

Calendar of Meets

Microcycle	Domestic	International	Location
8 (13 November)	■		World Cup
11 (4 December)		■	Christmas Meet
16 (8 January)		■	Atlanta
18 (22 January)	■		World Cups

Periodization

Training phase	Transition period	Building base	Intensity phase	Competition	Specific
Strength	Rehabilitation	Progressive	Power	Maintenance	Progressive/ Power
Endurance	Maintenance	General progressive	Mixed	Maintenance	Specific
Speed	Maintenance	Maintenance	Progressive	Maximum	Progressive

Testing Dates

Medical Control

Volume (scale 10–100 by microcycle 1–23)

Courtesy of Bob Bowman.

	March				April					May				June				July					August				
Week	24	25	26	27	28	29	30	31	32	33	34	35	36	37	38	39	40	41	42	43	44	45	46	47	48	49	50
Date	5 March	12 March	19 March	26 March	2 April	9 April	16 April	23 April	30 April	7 May	14 May	21 May	28 May	4 June	11 June	18 June	25 June	2 July	9 July	16 July	23 July	30 July	6 August	13 August	20 August	27 August	3 September
Event	All-Star			World Championships Trials							Ann Arbor				Local Seniors		Santa Clara				World Championships			U.S. Nationals			

Competition	Intensity	Specific	Peaking	T	Break
Maintenance	Progressive	Power	Maintenance	M	Break
Maintenance	Mixed	Specific	Maintenance	M	Break
Maximum	Maintenance/Progressive	Progressive	Maximum	M	Break

127

Table 9.2   Midseason Microcycle for Middle-Distance Swimmers

	Monday	Tuesday	Wednesday	Thursday	Friday	Saturday	Sunday
a.m.	Aerobic and technical work (kick and pull)	Power and overload training	Off	Aerobic and technical work (kick and pull)	Power and overload training	$\dot{V}O_2$max training	Off or easy swimming and recovery modalities
p.m.	Endurance training up to anaerobic threshold	Active rest (speed play)	$\dot{V}O_2$max training	Endurance and threshold training	Active rest and technical work	Off	

Table 9.3   Microcycle Examples

10- TO 11-SESSION MICROCYCLE

	Monday	Tuesday	Wednesday	Thursday	Friday	Saturday	Sunday
a.m.	X	X		X	X	X	(X)
p.m.	X	X	X	X	X		

9- TO 10-SESSION MICROCYLE

	Monday	Tuesday	Wednesday	Thursday	Friday	Saturday	Sunday
a.m.	X		X		X	X	(X)
p.m.	X	X	X	X	X		

m freestyle. However, Klete was given more aerobic mileage as is the normal course of action with a swimmer who is not particularly muscled and comes from a distance training background. Peter was given slightly less volume training because of his muscular build and the residual fatigue that can occur during sharpening training.

Mental Preparation

Mental preparation is the key to outstanding performance in world-level competition. This includes practicing visualization, building confidence, and learning not to worry about temporary setbacks. We encourage our athletes to visualize their races starting about 6 weeks prior to the competition. They learn to see the race unfold the way they want it to, and we encourage them to be as specific as possible in their mental imagery. The taper period is very valuable as a time to rehearse various aspects of the race. Although it is impossible to replicate the entire race in training, it is possible to focus on parts of the race. Swimming at specific speeds for splits of the race, practicing the start and race speed turns and finishes, is critical. We encourage our swimmers to visualize themselves swimming in the competition while performing these training tasks.

The coach must be the picture of confidence at this time. Each swimmer will go through days and sessions where he does not feel good or fast as the taper progresses. In our program we do not put a lot of emphasis on how each swimmer feels on a day-to-day basis. We focus on the stroke technique and times swum at certain distances. You do not have to feel good to swim fast.

Table 9.4 Michael Phelps' World Championship 2007 Taper Plan

Day, date	Days out	VOLUME (TRAINING TYPE)	
		a.m.	p.m.
Monday, 5	20	4.0 (P)	6.0 (E)
Tuesday, 6	19		6.0 (AR)
Wednesday, 7	18	3.5 (T)	5.0 (MVO$_2$)
Thursday, 8	17		6.0 (E)
Friday, 9	16	3.5 (P)	4.5 (T)
Saturday, 10	15	5.0 (MVO$_2$)	
Sunday, 11	14		4.0 (T)
Monday, 12	13	3.5 (P)	5.0 (E)
Tuesday, 13	12	4.0 (T)	Travel to Australia
Wednesday, 14	11		
Thursday, 15	10	Arrive (10:15)	3.5 (T)
Friday, 16	9	4.0 (T and P)	4.0 (E)
Saturday, 17	8		5.0 (AR)
Sunday, 18	7	3.5 (P)	3.5 (MVO$_2$)
Monday, 19	6		4.5 (T)
Tuesday, 20	5	3.0 (T and P)	3.0 (Pace)
Wednesday, 21	4		4.0 (T)
Thursday, 22	3	2.5 (T)	2.5 (T)
Friday, 23	2		3.0 (T and Pace)
Saturday, 24	1	2.5 (Easy)	

E = endurance training up to anaerobic threshold; MVO$_2$ = quality speed work at VO$_2$max or higher; AR = active rest (speed play or fartlek training); T = technical work (stroke improvement); P = power training (resistance work with equipment); Pace = rehearsal swimming at race speed.

Three days before the Melbourne World Championships, Michael had one of his worst workouts of the year. But he knows that how he feels 3 days before the competition is not important. He was well prepared for the year preceding the event, and one practice would not affect his performance in the meet. The athlete and coach must have faith in the plan and not panic when temporary setbacks occur.

The peaking period is not a magic time when coaches create miracles. The peaking period is a natural outgrowth of the entire training year. If the preparation during the hard training periods is well planned and executed, then the taper will be successful. The coach must "sell" his plan to the athletes and over a period of years refine it to meet the needs of the individual athletes. When the coach and athlete have confidence in each other and trust in the plan, then the stage is set for breakthrough performance.

Applying Bob Bowman's Expert Advice

As coach Bob Bowman indicates, failing to plan is planning to fail, and each athlete's specific characteristics should be kept in mind when designing a long-, medium-, and short-term training and peaking plan. In this section, readers can learn how Bob designs a tapering plan for swimming, keeping in consideration the specificities of the event and the mental preparation of the athletes. Key elements are keeping things in perspective no matter how a swimmer feels day-to-day (i.e., having faith in the plan despite temporary setbacks) and the concept that the taper is a natural outgrowth of the entire training year.

Martin Fiz

Triumphing in Eight Elite Marathons

On August 13, 1995, I won the marathon at the World Championships in Gothenburg, Sweden, with a time of 2:11:31. Two years later, on August 10, 1997, I finished second to Abel Antón at the World Championships in Athens, Greece. Many people ask me whether I would use the same training program in the lead-up to the 1997 event and the same race tactic during that marathon if I had the chance to race it again and aim for gold. In that marathon run, my main rival was Abel Antón, who was a fast runner in the final meters and ran very smoothly in flat courses. On the other hand, he had difficulties responding to strong surges and competing with athletes who maintained a constant, fast pace from the beginning of the race in an attempt to drop him before reaching the final kilometers of the marathon. Courses with a lot of turns were not his favorite either. In Athens, after I led the race, setting a fast pace for more than 19 km, Antón beat me by sprinting in the final 200 m. I was second.

More than 10 years have gone by, and looking back at every aspect of my preparation, I can assure you that I would repeat my training plan and race tactic. There would only be one change: I would look for an ally with characteristics similar to mine to help me set a pace to weaken my main rival. I am convinced that by training faster or by carrying out a taper including more anaerobic training sessions, I would not have improved my performance on the day. Marathon running requires a lot of hard pace running although there is more than one way to win.

MARTIN FIZ

Spain's Martin Fiz celebrates with the Basque (left) and Spanish flags after winning the marathon at the fifth World Track and Field Championships in Gothenborg, Sweden, in 1995.

© AP Photo/Thomas Kienzle

Martin Fiz is considered the world's best marathon runner of the 1990s. Martin took part in 16 elite marathon races between 1993 and 2000, winning eight of them with times ranging from 2:08:05 to 2:12:47. He won gold at the European Championships in 1994 and World Championships in 1995 and silver at World Championships in 1997.

Marathon Training

It takes special preparation to run a long-distance event in another country or continent, travelling to which requires the athlete to pass several time zones. The athlete will need to adapt his body clock to the new time zone and season of the year. We need to consider factors such as allergic reactions in the spring and climate conditions such as temperature and humidity. It is different to run in the afternoon or in the early morning: When do I eat lunch? How long will I need to digest my last meal? It is also important to know the characteristics of the race course: A flat course means a fast race whereas a turning course means slow race and surges. Athletes must visualize how the race of their life is going to be and how competitors are going to move. Athletes have to be alert, analyzing the virtues and weaknesses of potential rivals. Although there are many competitors, one must take a reduced group into consideration. From here, it is a matter of believing that one's goal is achievable and starting one's preparation.

To prepare for an event of the magnitude of the marathon at the World Championships, several factors that go beyond physical training itself need to be considered:

- Realistic goals
- Place, date, and time
- Race course
- Rivals
- Physical training

The aim of the taper in any training program is to achieve or optimize peak performance. But, the peaking starts even before training begins. An athlete must establish a realistic goal. On the day of competition, it is unwise for the athlete to aim for levels that have never been experienced; it would be disastrous to start running a marathon at a pace that the athlete's body is not familiar with. Following is my plan for the marathon run at World Championships.

First 4 Months

This phase of physical training is enjoyable for many athletes because anxiety and stress are absent and workouts can be quite varied. Training with a medicine ball will strengthen the upper body. Fartlek running (i.e., programmed pace changes) on undulating surfaces will strengthen the legs considerably. I can run more than 3 hours at a very slow pace, 5 min per kilometer. A lot of coaches and physiologists consider this "garbage mileage," but I disagree. In addition to preparing the muscles for the impacts they will suffer during the 42,195 m, these workouts allow the body to adapt to and experience the sensations of running without glycogen, relying on fat, the so-called low-quality fuel.

Four months of hard and intensive training, with daily double sessions, hours of physiotherapy, and a strict nutrition plan, will help an athlete arrive at the competition in peak fitness (table 9.5).

Finding the Perfect Pace

Once the initial strength phase has been completed, it is time to start the 3-week marathon-specific preparation. From now on, the key is to adapt to and memorize the pace. To do that, I run many days at marathon pace. The only problem is that I have to endure being tired through the day. The daily work with little rest and fast running

Table 9.5   Sample General and Marathon-Specific Preparation Weeks

GENERAL PREPARATION WEEK		
	a.m.	**p.m.**
Monday	10 km: 4 min/km Stretch Medicine ball exercises	12 km: 4 min/km
Tuesday	20 min jog + 30 min fartlek + 10 min jog	Rest
Wednesday	Stretch Massage	Rest
Thursday	10 km: 4 min/km Stretch	15 km: 4 min/km
Friday	30 min jog + 2 × 200 m 5% uphill, recovery jog downhill + 20 min jog	
Saturday	Rest	Rest
Sunday	20 km: 3:40 min/km	Rest

MARATHON-SPECIFIC PREPARATION WEEK		
	a.m.	**p.m.**
Monday	18 km: 3:40 min/km Stretch	12 km: 4 min/km + 6 km: 3:15 min/km
Tuesday	15 km easy jog Stretch Massage	4 km jog + 10 km: 3:20 min/km + 3 km slow jog
Wednesday	Rest	4 km jog + 3 × 5 km: 3:05 min/km/4 min passive recovery + 2 km: 4:30 min/km
Thursday	10 km: 4 min/km	15 km: 4 min/km
Friday	Rest	4 km jog + 25-30 × 400 m: 1:12 min/1:15 jog recovery
Saturday	12 km: 4 min/km	12 km: 4 min/km Massage
Sunday	5 km: 4 min/km + 10 km: 3:12 min/km + 5 km: 3:35 min/km + 5 km: 3:05 min/km	Rest

pace will make my muscles feel heavy and my mood gray. The biggest training load in terms of kilometers comes during the middle phase: 220 km weeks and an average of 180 km per week will mark the end of a preparation that seems perfect to me. The training load, recovery, and rest will determine the adaptation to training and the success of the campaign (table 9.5).

Tapering for the Marathon

The time to taper is here. As a marathon runner, I tried to rectify the tapering mistakes I made during this training phase when I was a 5,000 m runner. Here are my characteristics as a runner: 55.5 kg, 169 cm, maximum heart rate 188 beats/min, peak lactate after a progressive treadmill test (1% incline, 3 min increments) 8 mM, peak velocity 22 km/hr. These values suggest that it will be really difficult for me to beat my opponents in the final meters of a race no matter how much anaerobic training I do unless I come across another runner of similar characteristics. More anaerobic training in the final phase of preparation will only induce additional muscle trauma. In fact, when tapering to run the 5,000 m I suffered more injuries than I ever did preparing for the marathon. Muscle problems appeared quite often in my hamstrings, Achilles tendons, and calves, which forced me to slow down my preparation to recover muscle strength. On the other hand, when I was tapering for the marathon, minor muscular problems could be taken care of without having to stop my preparation. Tired and somewhat overstressed muscles can be easily relieved with proper massage and physiotherapy sessions. The main risk could come from physiological deficiencies, such as below-reference micronutrient, blood, and immunological values attributable to the high training volumes.

Therefore, my suggestion for runners training at continuous and fast pace is to taper in an effort to enhance their strength. For example, which one would you choose?

(A) 4 × 2,000 m at 2:40 min/km with 6 min recovery

(B) 8 × 2,000 m at 2:55 min/km with 2:30 min recovery

Theoretically, an athlete should be able to perform both training sets. He has the required quality, but which is more appropriate? In example A, the athlete will run 15 s faster than in example B, and this implies a higher leg stride cadence. He will feel fast and will be convinced that he is in top form. In example B, the athlete will train at a slower pace but will run twice as much and the recovery is much shorter. If the athlete chooses to perform this set, he will achieve a good pace, similar to that of competition; will memorize racing pace; and will gain endurance.

Which one would I choose? Without a doubt, I would choose B. Why? One reason is because example B provides the previously mentioned training effects. Also, the athlete who chooses A will train with an accumulation of lactic acid that will never be reached during a marathon race. This may lead to overpacing during competition. In fact, some athletes like to compete at a high level 2 weeks before the marathon. They take part in shorter distance racing (15 km to half marathon) to convince themselves of their fitness status. In my view, they make the mistake of giving too much effort at this point: They race with higher lactate values, and they take longer to recover. My advice is to race at the same pace intended for the day of the marathon. I thus recommend tapering by performing endurance-pace training sessions.

As training cycles are accomplished, the pace may be increased, but not too much, and the recovery times between repetitions during interval sessions or "pace-change" sessions can be reduced. During the first months of preparation, the pace will be slower and the recovery times longer, as shown in the box Interval Training Sets for Race Pacing and Tapering.

The duration of my perfect taper is 21 days. One week before, I include a race of 15 to 21 km. From then on, continuous running and pace intervals increase, but recovery time between repetitions decreases (see the sample interval training sets for tapering).

Interval Training Sets for Race Pacing and Tapering

Following are some examples of interval training sets I used in my preparation. They are not very different in terms of repetition number and distance but are quite different in terms of pace and recovery times.

Sample Interval Training Sets for Perfect Race Pacing

- 12 × 1,000 m: 3:10 min/km with 2:30 min recovery jog
- 25 × 400 m: 1:12 min/km with 1:15 min recovery jog
- 3 × 4,000 m: 3:12 min/km with 4 min recovery jog
- 5 × 3,000 m: 3:10 min/km with 4 min recovery jog

Sample Interval Training Sets for Tapering

- 3 × 4,000 m: 2:55 min/km with 3 min recovery jog
- 25 × 400 m: 1:10 min with 45 s recovery jog
- 15 × 1,000 m: 2:55 min/km with 1 min recovery jog
- 5 × 3,000 m: 2:57 min/km with 2 min recovery jog
- 70 min continuous run, increasing pace from 3:40 min/km to 3:10 min/km

The main difference between race pacing and tapering sets is the recovery times. During the taper, my trick is faster running than months before and shorter recovery times: Recovery times are key. In this way, the athlete reaches a fitness peak, gets to competition day fresh and "hungry," and has a well-established race pace in his brain and in his legs.

Using this method I achieved athletic success and extended my athletic career to the age of 37. That was my age when I finished sixth at the Sydney 2000 Olympic Games: Olympic Diploma. I think my athletic longevity is in part attributable to the fact that my training programs always focused on aerobic work. I am 46 years old now, I still take part in endurance races, and I even win some of them. I still perform the same type of taper to target a specific race, except that I only run one session per day now.

AT A GLANCE   Applying Martin Fiz's Expert Advice

In this section, Martin Fiz discloses his training methods to optimally prepare for a world-class marathon race, which start by using one's brains to analyze the race conditions and set realistic goals. Martin also provides weekly training plans for the different preparation phases, including general preparation, race-specific preparation, and tapering. Key ideas from Martin's contribution include setting an achievable goal and analyzing all the environmental variables that may affect training programs and race tactics; performing huge running volumes to improve running economy and prepare your body for such a grueling race; learning the ideal race pace; and avoiding racing during the taper at velocities much faster than marathon pace. Another key element of the section is Martin's approach to visualizing the race and analyzing the strengths and weaknesses of the main rivals. Here is a clear example of his approach: In his gold-winning World Championships race in 1995, Martin did not panic when one of his main rivals surged in the final kilometers to take the lead. Instead, Martin timed him for 1 km, realized that his rival would not be able to keep such a pace to the finish line, and maintained his own pace as planned. Sure enough, a few kilometers later his rival "blew up," and Martin overtook him and never looked back.

Luc Van Lierde

Becoming a World Champion Triathlete

Every athlete dreams about becoming a world champion one day, but every world champion has had a long journey to get there. Here is a brief description of my own journey and some of the tapering strategies I used as a triathlete.

Background

Before I became a triathlete I participated in different sports, including hockey, soccer, judo, roller skating, and swimming. I trained in swimming for about 10 years. My best performances were set in 1988 with a time of 2:21 min on 200 m breaststroke. At the age of 19 I decided to shift from swimming to triathlon, where I made rapid progress. After 18 months of training I participated in the Olympic-distance World Championship in Florida, where I finished fourth. Over the initial years of triathlon training I had several injuries that slowed my progression as a triathlete. Still, I persisted in my drive to develop as an elite triathlete, and in 1995 at the age of 25 I was second in both the European and World Championships in Olympic distance and long distance as well.

In 1996, I became the European champion over the Olympic distance, and I realized my dream to win the Ironman Triathlon in Hawaii. One year later I took the world record in long-distance triathlon, 7:50:27, a record that stands after more than 10 years. In 1999 I took my second win on the Ironman in Hawaii and became the world champion for the third time. After 1999 my career as a triathlete suffered. I was afflicted by a series of injuries, which seriously affected my motivation. However, in fall 2005 I decided to go for it one more time and began training to prepare for a final comeback. In October 2007, at the age of 38, this resulted in rank 8 at Ironman Hawaii. I continue working hard, and we will see where the story ends.

LUC VAN LIERDE

Luc Van Lierde is one of the most outstanding triathletes in the history of the sport. In 1996 he won gold at the European Championships and silver at World Championships in the Olympic distance. He also became the first-ever European to win Ironman World Championships in Hawaii. The following year he set the fastest-ever time in an Ironman event with 7:50:27, and in 1999 he won Ironman Hawaii for the second time in his career.

Luc Van Lierde of Brugge, Belgium, jubilates as he crosses the finish line to win the 1997 German Ironman triathlon competition. Van Lierde clocked a new official record time of 7:50:27.

© AP Photo/Frank Boxler

Tapering Procedures

Following are examples of tapering procedures that I have used to prepare for Olympic-distance and Ironman-distance triathlon races. These tapers have worked for me, but this does not mean they will be ideal for other athletes preparing for similar events.

Tapering for Olympic-Distance Triathlons

- Day –6: cycle–run transition training consisting of 60 km cycling of which the last 20 km are at race pace, followed by a 2 km run also at competition pace
- Day –5: low-intensity recovery training session
- Day –4: swim training session with intervals at race pace
- Day –3: low-intensity recovery training session
- Day –2: short, high-intensity running fartlek
- Day –1: low-intensity recovery training sessions

Tapering for Ironman Triathlon

- Day –8: last extensive endurance training session in cycling (5 hr)
- Day –6: 1 hr run at about race pace
- Days –5 and –4: short and easy training sessions
- Days –3, –2 and –1: 1 hr bike session, a 20 min run and a 1 km swim

Tapering

My training for an important competition ends with tapering. Tapering is essential to obtaining maximal benefits from intensive training—it is the final critical stage of the training process. Over the years I have learned that there are no rules for tapering because the responses to this training phase vary widely among individuals. Some individuals prefer a long tapering period, whereas others apply a shorter period. Each athlete must develop the individual tapering procedure that yields maximal performance benefit at target competitions.

Some athletes make a mistake by using the final training sessions in the approach to the event as a sort of test, which may result in excess energy consumption. I have always found that low-intensity training sessions during tapering are very fatiguing. This is less the case for high-intensity training sessions. Maybe my body is asking for intensive exercise stimuli. During tapering I also feel consistently tired. Even going up the stairs feels like heavy exercise; my legs feel empty and tired. However, maybe this just indicates that they are ready for a burst of energy and optimal performance. Curiously enough, once I get this feeling, I am ready to perform. Furthermore, the importance of adequate nutrition during tapering is evident. A carbohydrate-rich diet during the 4 days before competition is pivotal to performance optimization, because muscles need fully repleted glycogen stores for maximal endurance capacity.

I have always used different tapering processes for short versus long triathlons. In the approach to long-distance events, I reduce my training volume and intensity much more than for a short-distance triathlon. I have found that when preparing for a short distance triathlon, I need high-intensity exercise stimuli until the day before the event. Muscle tone in the approach to a short distance event apparently must remain higher than for an Ironman. A number of short exercise stimuli the days before a short distance event are recommended to obtain the speed and explosiveness needed in competition.

For the tapering procedure that I have often used to prepare an Olympic-distance triathlon and for an Ironman triathlon, see the box Tapering Procedures.

Applying Luc Van Lierde's Expert Advice   AT A GLANCE

I hope this information will be of some help to you. There are no secrets in training or tapering, rather some logical rules to which each individual must adhere. I wish you success in your training and competitions. Friendly greetings,

Luc Van Lierde "Ironman"

Luc Van Lierde has repeatedly shown that he knows how to prepare himself to perform optimally for triathlon races ranging in duration from less than 2 hr (endurance event) to about 8 hr (ultraendurance event). In this section, Luc shares his views on tapering and peaking and also provides examples of his own tapering programs for these events. Key elements in Luc's section are the ideas that the taper is a critical stage of the training process that will help an athlete maximize the benefits of intensive training; that the taper should be individualized to suit each athlete's characteristics and performance targets; and that the final sessions of the taper should not be considered as a sort of fitness test at the risk of inducing excessive fatigue.

chapter 10

Tapering for Sprint and Power Events

Although the scientific knowledge about tapering and peaking for sprint and power events is less complete than that regarding endurance events, some valuable information has emerged in the last few years that can contribute to better planning for events of this type. As we saw in chapter 3, several physiological variables that are important for optimal strength, power, and speed performance improve as the result of an appropriate tapering program. These include increased fast-twitch muscle fiber diameter, enhanced contractile performance, increased neural activation, increased glycolytic flux during maximal exercise, and hormonal changes that could facilitate anabolic processes. These physiological changes are associated with performance gains at the whole-body level, as shown by the performance gains reported in table 6.1 for activities such as 25 to 50 m swimming, single- and multiple-joint muscular strength, weightlifting, and vertical jump. In the following pages, world-class coaches and athletes involved in sprint and power events tell us about their tapering and peaking approaches for major competitions.

Mike McFarlane

Tapering to Win International Sprint Events

At various times in an athlete's long-term training plan we design and implement a program geared to a major championship, such as the World Championships, European Championships, or Olympic Games. Programs to address competition-specific and individual athletes' needs going into a championship vary for senior and junior athletes.

The program presented next is for a senior athlete whom I coached and worked with over a 5-year time span, preparing and tapering for the European Championships. It covers the last 4 to 6 weeks before the competitive season begins.

Typical Warm-Up Session

- 6 × stride warm up: stride the straight, jog the bends
- Drills 2 × 20 m with 10 m run-off back to back over 30 m
- High knees
- Kickouts

MIKE MCFARLANE

Mike McFarlane is an athletics performance sprint coach in the United Kingdom. Mike has been a team coach for UK athletics since 1992 and has worked with European Championships, World Championships, and Olympic Games teams. Sprint athletes coached by Mike (e.g., Dwain Chambers, Tony Jarrett, and John Regis, to name but a few) have won countless gold, silver, and bronze medals at all major international competitions. Previously, Mike was a development officer with the Duke of Edinburgh's award scheme and was awarded an Order of the British Empire for services to sport and youth in 2001.

Julian Golding (far right) anchors the British team for the bronze medal in the men's 4 × 100 m relay at the 1997 World Track and Field Championships in Athens. Golding is one of many athletes whom Mike McFarlane coached to win world-level medals in track. Golding also won gold medals in the 1998 Commonwealth Games, the European Championships, and the World Cup.

© AP Photo/Doug Mills

- Superman
- Inside, outside
- 1 × 80 m from run
- 5 hurdles out, do each drill two times, and then run off 10 m
- Over, over, over the hurdles
- Swings
- Down the middle
- Spins
- 1 × 80 m from run; then 15 min mobility
- Specific and individual warm-ups

Following are descriptions of a few workout sessions that would typically make up my early-season program, in no particular order. I would also include weight sessions after a few of those workouts, especially the shorter sessions like drills and blocks and some rhythm running sessions.

Typical Workouts Before Competitive Season Begins

- 5 × 150 m with lap walk: 16 s down to 15.5 s working on rhythm
- 4 × 110 m working on rhythm and change of pace 30–30–50 (build speed, hold speed, top speed) time last 50 m
- Blocks over 10 m working on movement using the timing gates; then 4 × 100 m with 10 m rolling with lap walk
- Drill session plyometrics and blocks over 10/20 m and shuttles

- 300–200–100 working on rhythm and holding form; a medicine ball circuit afterward
- 150 m with 2 min recovery, then 3 × 60 m with 12 min rest
- 4 × 20 m using the time gates, then 1 × 350 m going past 300 m in 33 s

Our schedule for a male 100 m sprinter for the 10 days before a championship is presented in table 10.1.

Tapering and peaking involve being prepared at the start of the season and making sure athletes are in good condition at the time of competition. Train wisely, race wisely. We are human and mistakes will be made; we must learn from them and be better prepared next time.

Table 10.1 Final 10 Days of Training Before a Major Championship for a Male 100 m Sprinter

Thursday	Friday	Saturday	Sunday	Monday	Tuesday	Wednesday
Arrive at holding camp Long warm-up in the afternoon	No training	Sprint drills and active hurdle drills 140–120–100 full recovery	Blocks working on reaction: 4 × 10 m; 3 × 20 m; 3 × 70 m flying working on rhythm Weights: cleans and jump squats; upper body light	No training, massage	Warm-up running drills 3 × 30 m sled runs, light 2 × 20 m no sled 2 × 90 m flying with 20 m roll 2 × 50 m rolling with 20 m roll Medicine ball circuit	Warm-up running drills Blocks: 2 × 20 m form 4 × 20 m timed 4 × 20 m reaction work 6 × 20 m sticks (quick feet) 4 × standing long jump

Thursday	Friday	Saturday	Sunday	Monday	Tuesday	Wednesday
Travel to championships in the morning Massage in the afternoon	2 × 40 m flying Blocks: 2 × 20 m 2 × 40 m flying 1 × 80 m change of pace	No training, relaxing	Race	Race		

Applying Mike McFarlane's Expert Advice

AT A GLANCE

This brief contribution by coach Mike McFarlane describes training elements for an elite sprinter during the 4 to 6 weeks prior to the beginning of the competitive season and the day-to-day preparation in the 10 days before a major competition. A key idea in Mike's section is that an athlete's training program should prepare an athlete for competition from the beginning of the season, not just during the taper.

Jason Queally

Training for World-Class Sprint Cycling

The 1,000 m cycling time trial, known as "the kilo" in many countries, is an individually timed track race where each rider is released from an electronic starting gate. The time trial is run with one rider on the track at a time. The fastest rider over the kilometer distance is declared the winner. The current world record for this event is 58.875 s and was ridden at altitude (where the air resistance is less than at sea level) by Frenchman Arnaud Tournant, who dominated the event along with countryman Florian Rousseau during the late 1990s. More recently, the kilo has been conquered by the British riders: I won the Olympic title at the Sydney 2000 games and Chris Hoy won at the Athens 2004 games.

In 1995 I took up cycling as a 25-year-old interested in triathlon. During these initial years I approached training and periodization with an endurance philosophy: road time trials of 16 to 100 km with a 6-week taper focusing on speed and speed endurance. I used this protocol until 1998, when I realized that after the fourth tapering week I was beginning to lose my speed on the track. So I reduced my taper from 6 to 4 weeks.

In 1998 at the World Championships I competed against Arnaud Tournant, who beat the rest of the kilo field by a significant margin at the age of 19. I decided from that day that speed was the key for me to maximize my kilo potential. After extensive research, I developed a training program that works for me. My aim is to lay the foundations of strength and power to develop track speed for the whole season, and during the last 4 weeks prior to the major competition I attempt to maintain this speed and extend the distance at which I can sustain this speed. Therefore, it is only in the last 4 weeks of a seasonal macrocycle that my tapering protocol begins.

The idea behind my tapering program is to maintain the speed I have developed over the previous weeks and months. To achieve this I use a few differing training protocols with sufficient recovery to extend the distance at which I can sustain this speed. That means that the training and recovery during the taper are very similar to my training and recovery during my specific core training, the only real difference being an addi-

JASON QUEALLY

Jason Queally is one of the fastest sprint track cyclists in the world. Jason became Olympic Champion in the 1 km time trial in Sydney 2000, where he also took silver in the team sprint with Great Britain. Since 1999, Jason has won one gold, three silver, and four bronze medals at the World Track Championships. Jason also represented Great Britain at the 2004 Athens Olympics riding the fastest time ever as third rider in the team sprint.

Britain's Jason Queally waves after receiving the gold medal in the 1,000 m track cycling event at the 2000 Olympic Games in Sydney, Australia.

© AP Photo/Lionnel Cironeau

tional training day, with the training and recovery philosophy being identical: maximal efforts followed by full recovery.

I have to be fully recovered and very fresh going into this taper because it is more intense than my specific core training cycles. When preparing for the Athens Olympics in 2004, I took 5 or 6 consecutive days off from training during the intermediate point of my taper because I believed I was not recovering properly. Following this rest, when I felt recovered (I discuss later how I determine whether I am fully recovered) I continued with my cycles and went on to break the world record for the last lap of the team sprint.

Tapering Protocols and Cycles

The 3-day cycle begins with what I term the "neural component" of my training, which is the key to all my speed training. These are maximal efforts of 5 to 18 s duration (maximal power 2,200 W; power will fall from 2,200 to 900 W over this time). This initial day is usually followed by what I term "neural–metabolic sets." These are maximal efforts of 30 to 33 s (maximal power 2,200 W falling to 700 W). On the third and final day I do what I call "metabolic training." This consists of interval-type efforts and flying 1 km segments where the efforts are gauged but bring about absolute fatigue on completion (maximal average power for intervals 800 W for 120 s and for the flying kilo 800-900 W for 60 s).

Following are the two protocols I follow for the first 3 days of the cycle:

- **Day 1.** Flying 100 and 200 m efforts × 2, followed by 2 to 3 standing-start quarter laps
- **Day 2.** Standing-start 500 m × 3
- **Day 3.** Track intervals and flying kilos: (1) flying 500 m followed by rolling 750 m recovery × 4 repetitions per set, 2 sets per session; (2) flying 1 km × 2

or

- **Day 1.** Flying 100 and 200 m × 2, followed by two motorbike paced efforts of one lap
- **Day 2.** Standing-start 500 m × 3
- **Day 3.** Turbo intervals; two types. Interval type 1: 30 s of 800 W followed by 60 s recovery repeated 4 times per set, 2 sets per training session. Interval type 2: 20 s of 650 to 700 W followed by 10 s recovery repeated 7 times per set, 2 sets per training session. (I find turbo intervals more effective in generating fatigue, because with track intervals during the recovery phase you have to keep pedaling and focus on what is going on; with the turbo, after each effort you can stop pedaling and recover properly.)

I learned one of the interval training protocols from a general exercise physiology textbook (30 s on, 60 s recovery for 4 repetitions) and the other from a sport science publication by Tabata and colleagues (1997). I only used the first type of interval during my taper for the Sydney 2000 preparation.

My protocol for this cycle has now changed somewhat, although the pattern and ideas behind it are virtually the same. The only real difference is the exclusion of flying kilos and the inclusion of another type of interval (both intervals were a part of my specific training protocols, although they were only used occasionally in 2000 but not as part of my tapering protocols). Table 10.2 presents the cycle I followed for the tapering phase of my training for the Sydney 2000 Olympics.

Following are a few of my thoughts on some other aspects of training that may be important to athletes who want to perform to their best ability when it really counts.

Fatigue and Recovery

Recovery from training is just as important as the training itself and is crucial during the taper. I tend to train mainly in 2- or sometimes 3-day blocks on the track followed by 2 days of recovery. Because my training is highly neural (aimed at producing maximal

Table 10.2   Training Protocol and Cycles During the Tapering Phase Prior to the 2000 Sydney Olympics

Date	Venue	Track type	Training protocol	Gear ratio	Duration of Effort, s
19 August 2000	Brisbane	Concrete 333 m	Flying 200 m × 2	49 × 14 = 94 in.	10.94, 10.92
20 August 2000	Brisbane	Concrete 333 m	Standing-start 500 m × 3	50 × 14 = 96 in.	33.6, 32.7, 32.3
21 August 2000	Brisbane	Concrete 333 m	Flying 500 m intervals × 2, 4 reps per set (recovery between each rep 500 m at 60 s)	49 × 14 = 94 in.	32-33
22 August 2000	Brisbane		Rest day		
23 August 2000	Brisbane		Rest day		
24 August 2000	Brisbane	Concrete 333 m	Flying 100 m × 1, 200 m × 1 and 3 standing-start quarter laps	49 × 14 = 94 in. (flying), 50 × 14 = 96 in. (standing)	5.36, 10.64, 7.40, 7.28, 7.38
25 August 2000			Traveled to Melbourne		
26 August 2000	Melbourne	Wood 250 m	Standing start 500 m × 3	51 × 14 = 98 in.	32-33
27 August 2000	Melbourne	Wood 250 m	Flying 500 m intervals × 2, 4 reps per set	48 × 14 = 91.8 in., 49 × 14 = 94 in.	No times
28 August 2000	Melbourne		Rest day		
29 August 2000	Melbourne		Rest day		
30 August 2000	Melbourne	Wood 250 m	Flying 100 m × 2 and 3 standing-start quarter laps	51 × 14 = 98 in.	5.37, 5.22; no times for starts
31 August 2000	Melbourne	Wood 250 m	Flying 100 m × 1, flying 1,000 m × 2	51 × 14 = 98 in.	5.17, 59.2, 58.8
01 September 2000	Melbourne	Wood 250 m	Flying 500 m intervals × 2, 4 reps per set (recovery between each rep 500 m at 60 s)	51 × 14 = 98 in.	30-31
02 September 2000	Melbourne		Rest day		
03 September 2000	Melbourne		Rest day		
04 September 2000	Melbourne	Wood 250 m	Flying 100 m × 1, 200 m × 1	51 × 14 = 98 in.	5.08, 10.16 (personal best)
05 September 2000	Melbourne	Wood 250 m	Olympic sprint trials, 750 m	50 × 14 = 96 in.	45.9, 46.1
06 September 2000	Melbourne		Rest day		
07 September 2000	Melbourne		Rest day		

Date	Venue	Track type	Training protocol	Gear ratio	Duration of Effort, s
08 September 2000	Melbourne	Wood 250 m	Flying 100 m × 2, 2 × standing half laps (assessing some equipment that just arrived)	51 × 14 = 98 in.	5.08, 5.06, 11.4, 11.0
09 September 2000	Melbourne	Wood 250 m	Flying 1,000 m × 1	51 × 14 = 98 in.	58.5
10 September 2000	Melbourne		Rest day		
11 September 2000	Melbourne		Rest day		
12 September 2000	Melbourne	Wood 250 m	Flying 100 m × 1, flying 200 m × 1, standing 1/4 × 4	51 × 14 = 98 in.	No times
13 September 2000			Traveled to Sydney		
14 September 2000	Sydney		Rest day		
15 September 2000	Sydney	Wood 250 m	Flying 100 m × 1, standing 1/4 × 4 (from gate)	51 × 14 = 98 in.	No times
16 September 2000	Sydney	Wood 250 m	Olympic 1,000 m time trial	51 × 14 = 98 in.	61.609

levels of torque and power), anything beyond the 3-day window (determined using electronic timing and SRM power meter data) compromises my ability to produce these maximal forces on subsequent days. Thus I will not train when my ability to produce the forces described here is compromised through fatigue or illness.

Objective data are very helpful in identifying fatigue, but because I virtually always train maximally, I use subjective measures to gauge my absolute levels of fatigue. I believe there is a highly significant relationship between my general mood and fatigue from training. If I wake up in the morning of a training session feeling like I do not want to train, then I will not train. If I am becoming generally tired and grumpy, this is usually a sign I am becoming fatigued, and if this coincides with an increase in training times or a decrease in power, I will not train. Instead I take a break until my general mood is better and I feel like I want to train. When I start to get grumpy and tired, I may take several days off the bike, and during my specific core training phases I usually go into a road block rather than return to track training. If this occurs during the tapering phase, as it did in the lead-up to the Athens 2004 Olympics I have to return to the track because the road is not an option at this time.

On the road my approach is slightly different. I go into a road-training block (hoping to improve aerobic capacity) only after I feel fully recovered from the track cycle I have done previously. This may mean a break of 4 to 5 days. The subsequent road block will be 4 to 7 days of riding for 2 to 3 hr every day. I will train when I feel fatigued (grumpy) on these days. I am still able to produce effective powers on the road while still subjectively tired or grumpy. Once this cycle has finished, I will return to the track only when I feel fully recovered.

During my time as a cyclist I have experimented with various training, recovery, and tapering protocols, and I have found what best suits me to be competitive at the current world best level. This is not to say that my current protocols are ideal for me or for anyone else who follows them—they just currently work for me!

Of Cakes and Training Protocols

The analogy I usually give for developing a training protocol is that it's like making your favorite cake (chocolate in my case). To get your perfect cake, you have to make sure you have the correct ingredients. It is not adequate to just mix the ingredients together any old way and place them in the oven. You have to know the correct quantities of each ingredient, the order in which they are mixed, and how much time you spend mixing. I see this phase of the cake making as comparable to your specific core training, the day-in, day-out work you have to do. You may have problems in this stage, the core phase of your training (specific core training). I can give you the ingredients and quantities (training protocols), but the timing (the amount of training vs. recovery) is up to you; you have to find out what works for you. It will be a process of trial and error; you have to understand how your mind and body work if you are to get the most out of any training protocol I could give you.

Once you have identified the ingredients, quantities, and timing and have mixed everything up, you can place the cake in the oven at the ideal temperature for the ideal amount of time, and when you open those oven doors you have your perfect cake. That is the peaking–tapering phase.

AT A GLANCE   **Applying Jason Queally's Expert Advice**

I love food and eat what I enjoy. I have no special diet but ensure I have enough calories to train and recover effectively. If I feel hungry, I will eat. The only specific dietary requirement I try to uphold is taking a carbohydrate–protein shake after all my track sessions and sometimes after my road rides.

Jason Queally

In this section, Jason Queally describes his training methods for sprint track cycling. Jason has coached himself since the beginning of his cycling career, and his training methods are an interesting mix between the trial-and-error and the scientific approaches; between subjective and objective evaluation of his fitness and performance status. The way Jason organizes his training program is now known as "reversed periodization," through which speed and power are developed early in the preparation program followed by a focus on extending the time that speed can be sustained. Key concepts from Jason's section are developing strength, power, and speed for the whole season, and then working on maintaining the speed for longer during the taper; performing maximal efforts followed by complete recovery on the track; listening to your body and your mood to know when you are properly recovered and ready to perform your key training sessions; and understanding that training protocols are individual and that adaptation profiles change over time (i.e., a training and tapering protocol that works for Jason is not necessarily ideal for another athlete, and a protocol that has worked for Jason in the past may not be ideal for him in the future).

Shannon Rollason

Breaking Records in Sprint Swimming

In the lead-up to a major event, I like to taper my sprint athletes for 21 days before their first day of competition. This may not be their most important race, but they must feel good about the way they are swimming in those last few days prior to their major race.

SHANNON ROLLASON

Shannon Rollason is the head coach of the Australian Institute of Sport swimming program. Shannon started his coaching career in his early 20s and in just 10 years has become a reference in the swimming world. Between 2003 and 2007, swimmers coached by Shannon won four gold medals at the Olympic Games and a total of 11 gold, silver, and bronze medals at World Championships, breaking three world and Olympic records along the way. Jodie Henry and Alice Mills are two of the most successful swimmers coached by Shannon. Four of the swimmers he coaches represented Australia at the 2008 Beijing Olympics. Among other coaching distinctions, Shannon received the Australian swimming coach of the year award in 2004 and the Australian Swimming Coaches and Teachers Association coach of the year award in 2005.

The Australian 4 × 100 m freestyle relay team shows its gold medals after winning at the 2004 Olympic Games in Athens, Greece. The team set a new world record of 3:35.94. From left to right are Jodie Henry, Lisbeth Lenton, Petria Thomas, and Alice Mills.

© AP photo/Anja Niedringhaus

The two taper modalities that I have used most often with my swimmers are the linear and the exponential taper. Both Alice Mills and Jodie Henry have used these two types of taper with international success. With the exponential taper, there is a bigger volume reduction, approximately 68% of season preparation swimming distance in the first week, 48% in the second, and 33% in the third. My linear taper reduces to about 75%, 50%, and 25% over the 3 weeks.

In the last 21 days before competition starts, it is important that you and your swimmers are not trying to find speed; if you let it, it will find you (or them). The athletes should swim submaximal and easy sessions (easy–fast, I like to call it), without over-emphasizing the kick. This, along with stroke length, is what I look for. The other tool I use to assess swimmers is a 1×200 m heart rate check. I get the swimmers to swim at 30 to 40 beats below their maximum heart rate, and I monitor this against their time. It gives me an easy indication of where they are during the taper. At no time during the taper do I talk about fitness; generally I just emphasize efficiency and balance (between legs and arms) in the water.

My decision on what taper to use with Alice and Jodie depended on how uninterrupted their preparation was. Table 10.3 shows Jodie Henry's 2004 pre–Olympic Games taper.

The taper was a linear model. Her maximum volume in the 15-week preparation was 48 km, and her weekly average was 38.5 km. Six weeks out, Jodie became ill, and she only swam 15 km that week. That was the only significant hiccup in what otherwise was a good preparation, on the back of another good preparation for the Olympic trials, in which she swam 24.92 s in the 50 m, 53.77 s in the 100 m, and 1:59.26 in the 200 m, all

Psychology Is Crucial

Each swimmer has her own individual personality, and coaches should keep this in mind, especially in the last 5 to 7 days leading up to the competition and also in terms of rooming with other swimmers. In my experience, two swimmers who get highly excitable are not a good match. Another issue over an 8-day major meet is to match swimmers according to when they finish competing (e.g., swimmer A races on days 1, 2, and 4, whereas swimmer B competes on days 4, 7, and 8; this would not be a desirable match). The last thing I try to keep in mind is the emotional cost, especially with 50 m swimmers who race over the last 2 to 3 days. Performance can be compromised if swimmers have been overstimulated by fast swimming by their peers during the meet.

In the lead-up to the 2004 Olympic games, Jodie Henry was not the favorite of most people. I kept her name out of the newspapers as much as possible. However, after Jodie's split of 52.95 s in the 4 × 100 m freestyle relay, there was a lot of talk among swimmers that she was the woman who would win. Jodie got off track a little, and as her coach I was there to put her back, reminding her to stay in the moment—that what will be, will be! Her race instructions were nothing about the race itself, just the 10 to 15 min leading up to the gun—what to do on pool deck. To her credit, she followed these instructions to a tee. This was not the first time I had used this tactic; in 2003 at the World Championships I used the same approach with a young Alice Mills, who ended up winning two individual silver medals. Too many swimmers swim their race before the gun goes off.

I didn't talk to Jodie about what the opposition would do, like go out in under 26 s, because they didn't! In coaching, what you don't say is just as important as what you do say. And remember that timing is everything; don't say the right thing at the wrong time!

Table 10.3 Jodie Henry's Pre-Olympic Taper, 2004

WEEK 1							
	Saturday 24 July	Sunday 25 July	Monday 26 July	Tuesday 27 July	Wednesday 28 July	Thursday 29 July	Friday 30 July
a.m.	Gym 1,500 m: short power work in water	No training	4,500 m: pull set	4,400 m: aerobic pull and kick Gym	No training	600 m: light swim	No training
p.m.	No training	No training	4,000 m: HRC 2:08.80 (168 beats/min); 1,000 m decreasing set: 400, 300, 200, 100, lactate 6.8 mM	No training	No training	No training	4,200 m: HRC 2:11.80 (152 beats/min); session aerobic pull and kick
Comments					Travel to Germany for staging camp		As planned

	Saturday 31 July	Sunday 1 August	Monday 2 August	Tuesday 3 August	Wednesday 4 August	Thursday 5 August	Friday 6 August
a.m.	4,000 m: kick 1,000 m Gym	3,600 m: general light swimming	3,000 m: quality 4 × 50 on 55 s, push (27.3 s, SR 48; 28.4 s, SR 47; 28.5 s, SR 46; 28.0 s, SR 48)	No training	3,200 m: some quality kick: 25 m 15.1 s; 50 m 35.7 s	No training	2,500 m: very light general work
p.m.	4,000 m: 1,000 m HR 40 BBM: 2 × (5 × 100 on 1:50)/60 s rest, lactate after set 1, 2.7 mM; after set 2, 3.9 mM	No training	No training	3,000 m: general, some relay changeover work	No training	No training	3,500 m: HRC 2:06.70 (180 beats/min); 2 × 25 push band only 13.17 s, 13.01 s; 25 m kick 15.31 s; 25 m dive 12.14, SR 49; 2 × 30 m push 14.71 s, 14.40 s
Comments			Jodie didn't feel good, so 4 × 50 only once			Jodie sick, sore throat, so no training at all today	Jodie feeling better so did HRC; looked very good, so decided to do some quality training

	Saturday 7 August	Sunday 8 August	Monday 9 August	Tuesday 10 August	Wednesday 11 August	Thursday 12 August	Friday 13 August
a.m.	2,000 m: mix general Gym	400 m: choice light swimming	2,200 m: pull work, aerobic type	3,300 m: HRC 2:11.1 (166 beats/min)	No training	2,000 m	"Peanut pool" swim
p.m.	3,000 m: 750 m: 3 × 100 m decreasing on 1:50/60 s recovery (69.4 s, 65.3 s, 63.1 s, HR 164 beats/min); 3 × 100 m decreasing on 1:50/40 s recovery 65.8 s, 65.2 s, 63.3 s, HR 172 beats/min); 100 m at HR 10 BBM on 1:30 (58.4 s); 50 m at HR 10 BBM (28.4 s, lactate 8.3 mM)	No training	Last gym session	1,200 m: some "hypoxic" work: 4 × 50 pull, no breathing	2,300 m: quality: 50 m max pull 27.6 s; 2 × 50 in race suit, 1st dive (25.38 s, SR 55-51, 39 strokes), 2nd push (25.51 s, SR 53)/60 s recovery	1,000 m	No training
Comments		Arrived in Olympic Village			As planned. Traveled to training venue to get some good water: Ready to go!!		Did not attend opening ceremony

HR = heart rate; BBM = beats below maximum; SR = stroke rate; HRC = heart rate check.

Table 10.4 Jodie Henry's Olympic Week, 2004

WEEK 1							
	Saturday 14 August	**Sunday 15 August**	**Monday 16 August**	**Tuesday 17 August**	**Wednesday 18 August**	**Thursday 19 August**	**Friday 20 August**
a.m.	Women's 4 × 100 m freestyle relay (heat): split 54.04 s	No training	1,800 m: 3 × 100 m decreasing and some short sprints	No training	100 freestyle (heat): 55.19 s (2nd lap 27.97 s)	No training	No training
p.m.	Women's 4 × 100 m freestyle relay (final): split 52.95 s (2nd lap 27.38 s)	2,000 m: light and general	1,500 m: similar session but no 3 × 100 m. 27.46 s at 100 m SR, push	1,000 m: 25 m pace work, push	100 freestyle (semifinal): 53.52 s (2nd lap 27.28 s)	100 freestyle (final): 53.84 s (2nd lap 27.46 s)	Light swim after yesterday's excitement
Comments	**Won gold medal, set world record** Her fastest-ever 2nd lap, previous best 27.70 s: goal achieved				a.m., qualified 7th, only looked OK p.m., warmed up very well and was very fast without pushing it **Set World record**	Not as relaxed as night before. Also not as quick in warm-up **Won gold medal**	Did not compete in women's 4 × 100 m medley relay heat (Alice Mills swam freestyle leg in heat: 53.54 s)

WEEK 2	
	Saturday 21 August
a.m.	Race warm-up–style swim
p.m.	Women's 4 × 100 m medley relay (final): anchor leg 52.97 s
Comments	**Won gold medal, set world record**

Olympic total
3 gold medals
3 world records

HR = heart rate; SR = stroke rate.

personal best times. It is worth mentioning that the goal for her Olympic preparation, as far as I was concerned, was to improve her already strong second 50 m.

Jodie's swimming over the taper was 31 km in week 1, 20 km in week 2, and 12 km prior to day 1 of competition, women's 4 × 100 m freestyle relay.

What swimmers do during the major competition may be as important as what they do during the taper leading up to the event. Table 10.4 shows Jodie's racing and training schedule during the Olympic Games.

Applying Shannon Rollason's Expert Advice

As shown in the preceding section, Shannon Rollason has a thorough understanding of the different tapering options at hand and very clear ideas about the tapering strategies he likes to use with the swimmers he coaches. Shannon provides the readers with a day-to-day description of the taper implemented by Jodie Henry, which led her to win three Olympic gold medals, along with corresponding world and Olympic records. Key messages from Shannon's contribution are tapering for the first day of racing and then maintaining that performance level for the duration of the championships (this is in contrast with the strategy of trying to achieve peak performance at the end of a tournament, an alternative option that is sometimes used in the team sports); focusing on swimming efficiency and leg and arm balance in the water, rather than fitness and speed; considering each athlete's personality when trying to help her achieve optimal psychological preparedness; and using the right messages at the right time (i.e., what the coach says is as important as what he does not say, and he must avoid saying the right thing at the wrong time).

Vladimir Vatkin

Peaking for Men's Artistic Gymnastics

As in many other sports, preparation in gymnastics initially involves a general preparation phase, which consists of all-around physical preparation of the athletes. The specific preparation phase requires building routines that address the current code of points used in judging the competitions for which the athletes are training. The final phase of the preparation program involves direct preparation for a major competition.

Direct preparation should last about 3 weeks, plus a week for acclimatization when we travel overseas. Of these 3 weeks, the first two are the hardest ones, featuring the biggest routines and psychological loads (tests). The last week is a bit easier, with more

VLADIMIR VATKIN

Vladimir Vatkin is Australia's men's artistic gymnastics national coach and head coach of the Australian Institute of Sport men's gymnastics program. Vladimir started his coaching career more than 30 years ago in the USSR and Belarus. He coached Ivan Ivankov to the European Championships all-around gold in 1994 and 1996 and to the World Championships all-around gold in 1994 and 1997. Under Vladimir's supervision, Australian gymnast Philippe Rizzo won silver in the high bar at the 2001 World Championships and then gold in 2006. Vladimir received the 1994 coach of the year award in Belarus, all sports included, and in Australia he has been selected men's gymnastics coach of the year half a dozen times.

Joshua Jeffries, gold medal winner in men's all-around gymnastics, celebrates with his coach, Vladimir Vatkin, after his championship performance at the 2006 Commonwealth Games in Melbourne, Australia.

© AP Photo/Tony Feder

quality and less repetitions, but is still difficult enough. The week before the competition is the most difficult one, because gymnasts are often experiencing jet lag and still need to produce routines on the new equipment of the competition venue. This means that gymnasts usually cannot afford an easier time before competitions, because we need to show at the competition the same things we are doing in training.

Imperfect Preparation for Perfect Gymnastics

Like any other national coach, I had a lot of good competitions after good preparation, but several competitions were far from ideal. In 1994, at the World Championships in Brisbane for all-around final, I decided to take a public train with my gymnasts instead of the shuttle bus to get to the venue at what I considered to be a better time. Unfortunately, we missed the train connection, got lost, and were so late we thought we were going to miss the competition. We were very lucky that people picked us up on the highway and gave us a lift! We arrived at the competition at the very last moment, and the relief to get there was so big that the pressure of this major competition had gone completely. As a result, Ivan had his best competition ever. He nailed all six apparatus and won the competition, although he competed in the weakest group. All the strongest gymnasts competed after him, which is a major advantage in gymnastics.

In 2006, Philippe Rizzo's preparation for World Championships was really uneven; he went through two surgeries and did not have time for consistent training. We constructed a new high bar routine a couple of weeks before the last camp, and Phil went through the complete routine only three or four times during the camp (table 10.5). He was not very fit and was even thinking of not going to the World Championships. It did not get any better in Denmark: He was not able to catch the major release (which he had been doing for 8 years). The week we were training in Denmark he caught it for the first time in the team competition and qualified with the highest result. Things got worse after this: The high bar final was in a week's time but Phil's back became so sore that he did not train at all for 3 days and could not do very much. I was 90% sure I was going to withdraw him. His back got a little better 2 days before the competition but still was not good enough for normal training. On the day of the competition, it was the first time in the week he checked all his skills, and he made his best-ever competition routine: He won.

Conversely, Phil was very well prepared for World Championships 2002 and 2003 but only came sixth and fourth, respectively, after making the smallest mistakes. He was at a perfect peak of fitness at the Athens Olympic Games but failed to qualify for the finals. So at his worst physical level, in 2006, he managed to win. My concluding thoughts are that the efforts he made to overcome physical problems washed away any aspects relating to the psychological stress of competition.

AT A GLANCE   ## Applying Vladimir Vatkin's Expert Advice

In this section Vladimir Vatkin shows that elite athletes can come up with amazing performances in the most unfavorable circumstances and that everything is not always perfect behind a great performance, at least in artistic gymnastics. Key ideas to retain from Vladimir's section are that in contrast with other sports and preparation methods we have previously discussed, the final preparation phase in gymnastics features very heavy training loads, implying that a taper as such is not necessarily used to prepare for major competitions; that gymnasts will not normally display in competition performance levels they have not previously displayed in training (despite some happy exceptions, such as Philippe Rizzo's 2006 World Championships), so their competitive fitness should be similar to their training fitness; and that in a technical sport like artistic gymnastics, psychological preparedness may have a much bigger impact on competition outcome than optimal physical fitness.

Table 10.5   World Championships Direct Training and Preparation Schedule

AUSTRALIAN INSTITUTE OF SPORT/GYMNASTICS AUSTRALIA TRAINING CAMP		
Monday		Recovery
	10:00	Skinfold
	12:00	Meeting for all
Tuesday	8:30-11:45	Leg strength, strength, stretching, basics, drills and corner parts on floor, trampoline
	16:00-18:30	General warm-up, acrobatic warm-up, strength skills on rings Trampoline, apparatus skills and combinations
Wednesday	10:30-13:00	Leg strength, strength, stretching, basics, drills and corner parts on floor, trampoline Apparatus skills and combination Dismounts
Thursday	8:30-11:45	General warm-up, acrobatic warm-up, strength skills on rings, leg strength Routines*
	16:00-18:30	Strength, stretching, basics, drills and corner parts on floor, trampoline Apparatus skills and combination
Friday	8:30-11:45	Leg strength, strength, stretching Apparatus skills, combination and routines*
	16:00-18:30	General warm-up, acrobatic warm-up, strength skills on rings
Saturday	8:00-9:00	Leg strength, strength, stretching, basics, drills and corner parts on floor, trampoline
	10:30-13:30	General warm-up, acrobatic warm-up, strength skills on rings Parts and routines
Sunday		Rest, recovery
Monday	8:30-11:45	Leg strength, strength, stretching Apparatus skills and combination Dismounts
	16:00-18:30	General warm-up, acrobatic warm-up, strength skills on rings Apparatus skills and combination
Tuesday	8:30-9:55	Leg strength, preparation for the test
	10:00	Team test**
	16:00-18:30	General warm-up, acrobatic warm-up, strength skills on rings Mistake fixing, apparatus skills and combination
Wednesday	10:30-13:00	Leg strength, strength, stretching, basics, drills and corner parts on floor, trampoline Apparatus skills and combination Dismounts
Thursday	8:30-9:55	Leg strength, warm-up for the routines
	10:00	*One-touch warm-up—routine***
	16:00-18:30	General warm-up, acrobatic warm-up, strength skills on rings Mistake fixing, apparatus skills and combination

> *continued*

Table 10.5 > *continued*

AUSTRALIAN INSTITUTE OF SPORT/GYMNASTICS AUSTRALIA TRAINING CAMP		
Friday	8:30-11:45	Leg strength, strength, stretching Apparatus skills and combination
	16:00-18:30	General warm-up, acrobatic warm-up, strength skills on rings Apparatus skills and combination
Saturday	8:00-9:00	Leg strength, strength, stretching, basics, drills and corner parts on floor, trampoline
	10:30-13:00	Parts and routines
Sunday		Rest, recovery
Monday	8:30-11:45	Leg strength, strength, stretching Apparatus skills and combination Dismounts
	16:00-18:30	General warm-up, acrobatic warm-up, strength skills on rings Apparatus skills and combination
Tuesday	8:30-9:55	Leg strength, preparation for the test
	10:00	Team test**
	16:00-18:30	General warm-up, acrobatic warm-up, strength skills on rings Mistake fixing, apparatus skills and combination
Wednesday	10:30-13:00	General warm-up, acrobatic warm-up, strength skills on rings Apparatus skills and combination Dismounts
Thursday	8:30-9:55	Leg strength, warm-up for the routines
	10:00	*One-touch warm-up—routine***
	16:00-18:30	General warm-up, acrobatic warm-up, strength skills on rings Mistake fixing, apparatus skills and combination
Friday	8:30-11:45	Leg strength, strength, stretching Apparatus skills and combination
	16:00-18:30	General warm-up, acrobatic warm-up, strength skills on rings Apparatus skills and combination
Saturday	8:00-9:00	Leg strength, strength, stretching, basics, drills and corner parts on floor, trampoline
	10:30-13:00	General warm-up, acrobatic warm-up, strength skills on rings Parts and routines
Sunday		Rest, recovery
Monday	8:30-11:45	Leg strength, strength, stretching Apparatus skills and combination Dismounts
	16:00-18:30	General warm-up, acrobatic warm-up, strength skills on rings Apparatus skills and combination
Tuesday	8:30-9:55	Leg strength, preparation for the test
	10:00	Team test**
	16:00-18:30	General warm-up, acrobatic warm-up, strength skills on rings Mistake fixing, apparatus skills and combination

AUSTRALIAN INSTITUTE OF SPORT/GYMNASTICS AUSTRALIA TRAINING CAMP		
Wednesday	10:30-13:00	General warm-up, acrobatic warm-up, strength skills on rings Apparatus skills and combination Dismounts
Thursday	8:30-11:45	Leg strength, general warm-up, acrobatic warm-up Routines*
	16:00-18:30	General warm-up, strength, stretching, strength skills on rings Apparatus skills and combination
Friday	8:30-11:45	Leg strength, strength, stretching Apparatus skills and combination
	16:00-18:30	General warm-up, acrobatic warm-up, strength skills on rings Apparatus skills and combination Packing up for World Championships
Saturday	Early morning	Flying

*Routines: gymnasts present their routines to the coaches; **team test: 1.5 hr general warm-up and then gymnasts present their routines in competition order to the judges; ***one-touch warm-up routine: same without the judges.

Gary Winckler

Producing Extraordinary Sprint Runners

For most coaches and the athletes they train, there is nothing higher on our wish list of accomplishments than to achieve the best performance at the most important competition of the year. I have always trained my track-and-field athletes to give their peak performance of the year at the major championship for their event. This might be a college conference championship meet or a World Championships or an Olympic final.

In my more than 30 years of coaching I have listened to many other coaches' ideas and tried many routes to achieve this goal with athletes in a variety of events. Most of my experience has been working with athletes in speed and power events ranging from short and long sprints to hurdling and jumping events. Next I discuss some of these routes and explain what has worked well for my athletes in trying to achieve optimal performance at the desired time of year.

Timing of the Taper

Early in my coaching career I was led to believe that peaking and tapering constituted a special phase in the training and competitive year where a different approach should be taken to make sure the athlete was ready for the big event. Some examples of protocols I was led to try included these:

- Delaying the highest intensity of training until the last 14 to 21 days prior to the big event
- Dramatically decreasing the volume of training and increasing the number of rest days 14 to 21 days prior to the big event
- Using new and different training means such as swimming or running in water while reducing the number of training sessions on the track
- Devoting a larger proportion of training time in the last 3 weeks of training to psychological skills training

GARY WINCKLER

Gary Winckler has been producing extraordinary athletes for more than 20 years. Thirteen athletes coached by Gary have reached the Olympic Games in the sprints, hurdles, or relays, and since 1992 athletes under Gary's guidance have appeared in every World Championships and Olympic Games, winning gold, silver, and bronze medals. Perdita Felicien and Michelle Finn are two of the gold medal winners coached by Gary. Gary has served as head coach of the U.S. World Championships team, has directed the USA Track & Field national coaching education program, and is a level III coaching instructor for the International Amateur Athletic Federation.

U.S. athlete Michelle Perry, left, crosses the finish line to win the gold medal in the women's 100 m hurdles final at the 2007 World Athletics Championships in Osaka, Japan. Canada's Perdita Felicien, right, took silver, and Sweden's Susanna Kallur, center, finished fourth. Both Felicien and Kallur have benefited from Gary Winckler's coaching at the University of Illinois.

© AP photo/Anja Niedringhaus

All of these schemes have value and can contribute to achieving the peaking and tapering desired. However, the common thread seen in these examples that I found did not contribute to the desired goal was that these tasks were only emphasized in the last 14 to 21 days before the major event. Emphasizing different schemes or exercise protocols in the 2 or 3 weeks before a major competition can interrupt the natural rhythm and flow of training and recovery that the athlete has become accustomed to during the course of training.

My experience has been that the process of peaking should begin on the very first day of training in the general preparation segment. The habits we establish and the baseline of conditioning and technical and tactical skills we ingrain in our athletes become the pedestal from which we launch our final peaking or tapering process at the end of the year. Early in the training year I present all of the skills that the athletes will need to have in their arsenal to approach the highest stress of the competitive year calmly and with confidence.

Thus, in terms of peaking and tapering, I simply work to help my athletes improve their critical skills over the course of the entire training year, working from very closed teaching environments to very open, competitive ones. My experience has been that as athletes achieve confidence in their skills, we can use this as a focal point in competition. In our sport athletes have almost total control over the performance, so if we remain focused on the skills that are most important to their performance, they will perform better.

Psychological Fitness

If you read the material I have contributed to this chapter, you can see that preparation for any high-level athletic event involves not only physical skills but psychological fitness as well. My approach to psychological fitness has largely been to help my athletes achieve a level of confidence in their physical skills and to know how to use them. Athletes who are confident about their ability to perform the required skills will be more focused and thus not overwhelmed by the high stress of major competition.

Final Taper

How do we approach the highest competition? During the early and middle part of the competitive cycle, we observe how athletes are performing in competition. We determine their most critical errors and why they are occurring. Using this information we select coping mechanisms, whether they are physical or psychological tasks, that we spend time developing in training and competition. As we approach the later stages of the competitive season and therefore the peak event of the year, we continually practice the familiar tasks that we believe are critical to the performance. Along with this comes a gradual decline in the volume of training and maintenance of high intensity work used throughout the entire training year. In sprinting and hurdling, for example, it is important to use high-intensity, competition-specific runs at least once every 3 to 4 days to maintain not only fitness but technical rhythm and competence as well. Similarly, we do not abandon strength training but continue with short, intense bouts of strength work right up to the day before competition. The key element here is to not introduce new exercises and training protocols but rather stay with the familiar and proven exercises and seek better timing and execution of those exercises. This method supports the psychological needs of the athlete and improves the critical physical skills required for the best performance.

Applying Gary Winckler's Expert Advice

AT A GLANCE

In this section by Gary Winckler, readers can learn about the peaking strategies used by an extremely experienced sprint coach, who discusses not just the physical aspects of the optimal preparation of his athletes but also technical and mental issues related to peak performance. Key ideas in Gary's section are planning a taper that does not interrupt the natural rhythm and flow of an athlete's usual training and recovery scheme; focusing on psychological fitness to help athletes be confident about their competence to perform the required skills under the pressure of competition; slightly reducing the training volume but maintaining similar intensity during the final taper; maintaining short, intense bouts of strength work right up to the day before competition; and performing the sessions at the time of the main event.

The results of the competition went as planned, except that we came up 0.03 s short in the final! The first round was a solid race with easy qualification to the semifinal. The semifinal was a very competitive race, and our start to the first hurdle was not as good as we had planned. Late-race technical proficiency saved the day. In the final, again the start was not as good as most of the field, but the midrace hurdling was the best in the field and produced the needed momentum to finish strongly off of the last hurdle to gain a silver medal.

Sample Taper

Table 10.6 provides a program that a 100 m hurdler used to achieve a season's best performance and a silver medal performance at the 2004 World Championships. You will notice a rhythm in the training prescription. The order of the workloads did not alter greatly, and roughly speaking the program was a very simple progression of the following elements:

- General strength and running
- Lactacid capacity runs and strength work
- Hurdle-specific runs
- Rest

Table 10.6   World Championships Taper Program for a 100 m Hurdler

WEEK 1		
Monday 13 August	Travel to training camp	
Tuesday 14 August	Travel to training camp	
Wednesday 15 August	800 warm-up: jump ABS circuit 5 5 × 10 low leg lifts 5 × 10 side-ups 5 × 10 back hypers with twist 5 × 10 rockers 5 × 10 crunch with twist 5 × 10 low circles	Tempo: 100s 3 × 4 × 100 tempo runs (50 m walk, 100 m walk) Hurdle mobility II 3 × 8 hurdle walk-overs 3 × 8 hurdle bent leg—middle of hurdle 3 × 8 hurdle over–unders 3 × 16 hurdle in-place dual trail legs 3 × 10 crocodile walk with pause
Thursday 16 August	2,400 warm-up: run 3 × 200 lactacid capacity (6′) ACAP 1/2 squat: high 2 × 60 1/2 squat (2′) 2 × 72.5 1/2 squat (2′) 2 × 85 1/2 squat (2′) 2 × 95 1/2 squat (2′) 8 × 17.5 jump squat Hang from bar: 2 × 30″ APOW step-up 2 × 4 × 30 step-up (3′) 2 × 4 × 37.5 step-up (3′) 10 × 4 × 42.5 step-up (3′)	Mobility: hurdle snake 6 × 6 one-step snake under hurdles lined up with cross-bars in a straight line and every other hurdle facing opposite direction 3 × 10 crocodile walk with pause 8 × 100 sprinter's cool-down Tempo crescendo runs Long controlled acceleration and deceleration Mobility hip 1 Mobility shoulder
Friday 17 August	1,300 warm-up: movement Hurdle mobility 3 × 8 hurdle walk-overs 3 × 8 hurdle bent leg—side of hurdle 3 × 8 hurdle over–unders 3 × 10 crocodile walk with pause 3 × 25 starts over 2 Hurdles 3 × 30 starts over 3 hurdles 1 × 8 hurdle at 8.20 m 1 × 7 hurdle at 8.20 m 1 × 6 hurdle at 8.20 m	Mobility: hurdle snake 6 × 6 one-step snake under hurdles lined up with cross-bars in a straight line and every other hurdle facing opposite direction 3 × 10 crocodile walk with pause 8 × 100 sprinter's cool-down Tempo crescendo runs Long controlled acceleration and deceleration Mobility hip 1 Mobility shoulder

Saturday 18 August	Rest	
Sunday 19 August	800 warm-up: dynamics ABS circuit 3 3 × 10 scissor legs; keep low to the ground 3 × 10 reverse hypers 3 × 10 low circles 3 × 10 rockers 3 × 10 Bart Connors 3 × 10 back hypers 3 × 10 low circles 3 × 10 V-ups 3 × 10 flutters 3 × 10 prone hip extension 3 × 10 medium circles 3 × 10 toe touch	Tempo: 100s 2 × 4 × 100 tempo runs (50 m walk, 100 m walk) Mobility: hurdle snake 6 × 6 one-step snake under hurdles lined up with cross-bars in a straight line and every other hurdle facing opposite direction 3 × 10 crocodile walk with pause

WEEK 2

Monday 20 August	800 warm-up: jump GCAP snatch: high 2 × 30 snatch from floor or hang (30″) 2 × 35 snatch from floor or hang (30″) 2 × 42.5 snatch from floor or hang (30″) 6 × 2 × 47.5 snatch from floor or hang (30″) Hurdle jumps 1 (balzi su ostacoli) 1 × 4 × 6 hurdle jumps—12″ hurdles spaced at 1.40 m (2′, 6′)	Lactacid: 60–60–60 2 × 180 in for 60 + 60 hold + 60 in ABS circuit 5 5 × 10 low leg lifts 5 × 10 side-ups 5 × 10 back hypers with twist 5 × 10 rockers 5 × 10 crunch with twist 5 × 10 low circles 8 × 100 sprinter's cool-down Tempo crescendo runs Long controlled acceleration and deceleration Mobility hip 1 Mobility shoulder
Tuesday 21 August	1,300 warm-up: movement 2 × 25 starts over 2 hurdles 2 × 30 starts over 3 hurdles 3 × 50 medium hurdle rhythm endurance 1 × 5 hurdle at 8.20 m 1 × 7 hurdle at 8.20 m 1 × 8 hurdle at 8.20 m	Mobility: hurdle snake 6 × 6 one-step snake under hurdles lined up with cross-bars in a straight line and every other hurdle facing opposite direction 3 × 10 crocodile walk with pause 8 × 100 sprinter's cool-down Tempo crescendo runs Long controlled acceleration and deceleration Mobility hip 1 Mobility shoulder
Wednesday 22 August	Travel to Osaka	
Thursday 23 August	1,300 warm-up: movement Tempo: 100s 3 × 4 × 100 tempo runs (50 m walk, 100 m walk)	Hurdle mobility II 3 × 8 hurdle walk-overs 3 × 8 hurdle bent leg—middle of hurdle 3 × 8 hurdle over–unders 3 × 16 hurdle in-place dual trail legs 3 × 10 crocodile walk with pause

> continued

Table 10.6 > *continued*

WEEK 2 *(continued)*		
Friday 24 August	1,700 warm-up: speed 2 × 25 starts over 2 hurdles 3 × 30 starts over 3 hurdles ABS circuit 4 2 × 10 V-ups 2 × 10 back hypers 2 × 10 prone hip extension 2 × 10 toe touch 2 × 10 double leg eagles 2 × 10 low-level bicycle 8 × 100 sprinter's cool-down	Tempo crescendo runs Long controlled acceleration and deceleration Mobility hip 1 Mobility shoulder
Saturday 25 August	Rest	
Sunday 26 August	800 Warm-up: jump GCAP snatch: high 2 × 30 snatch from floor or hang (30") 2 × 35 snatch from floor or hang (30") 2 × 42.5 snatch from floor or hang (30") 2 × 47.5 snatch from floor or hang (30") 8 × 100 sprinter's cool-down	Tempo crescendo runs Long controlled acceleration and deceleration Mobility hip 1 Mobility shoulder
WEEK 3		
Monday 27 August	1 × 100 competition 100 m hurdles (heat)	
Tuesday 28 August	1 × 100 competition 100 m hurdles (semifinal)	
Wednesday 29 August	1 × 100 competition 100 m hurdles (final) **Silver medal**	

ACAP: Alactacid capacity—Fast movement, low reps (2-4" effort), short to medium recovery consistent with ASSE (alactic short speed endurance running).

APOW: Alactacid power—Fast movement, medium reps (4-8" effort), recoveries consistent with speed development (90%).

GCAP: Glycolytic capacity—Fast movement, low reps (2-4" effort), short recovery consistent with GSSE (glycolytic short speed endurance running).

The volume of training performed at this time was slightly less than that used during most of the competitive training segment. Training intensity was the same as that used during the rest of the year. Most training sessions were carried out in the afternoon, except for a few morning sessions to match the time of the first round of competition. During the 21 days leading up to the major competition, the athlete and I discussed key technical points we wanted to focus on. All runs were timed and compared with results achieved in previous seasons. These timed runs all indicated that special fitness and technical preparations were as good as or better than those of any previous season.

Athletes should stay in their familiar training environment prior to competition. Training camps are all the rage with many teams and individuals leading up to a major competition. If these camps are not properly planned with familiar food, good sleeping arrangements, and good training support, the camps merely become another obstacle that the athlete needs to adapt to leading up to the major competition. Athletes should travel early to major events only when they need time to adapt to a time change.

chapter 11

Tapering for Precision Sports

In contrast with most individual sports, in which performance is often related to an athletes' physiological capacities and basic training variables, performance in precision sports is a measure of an athlete's accuracy in activities that do not rely on cyclic movements. Examples of these precision sports include shooting, archery, golf, and curling. Unfortunately, no scientific knowledge is available regarding optimal tapering and peaking strategies for these sports. Should an athlete shoot more arrows in the lead-up to the Olympic Games or less? Should a golfer predominantly concentrate on her drive or her putting skills in the days before a major tournament? How do strength and endurance training affect shooting accuracy? Until researchers answer these and other questions relating to tapering and peaking for precision sports, all we can do is hear about the personal experiences of some great champions.

Simon Fairweather

Journeying to Olympic Gold in Archery

The notion of the taper in archery is not often discussed either formally or informally among competitors and coaches. There is no academic research on the topic for the sport. The week leading into a competition is significantly influenced by the timetable, logistics, and weather conditions at each event, and an athlete's ability to adapt during final preparations is essential. Optimizing mental energy is just as important as peaking physical energy levels, if not more so. Getting it right for the 5 days of competition is quite personal for each individual and relies on many factors including the athlete's experience and the support he has.

During the 18 years I competed at the national and international levels (1987-2005), I had two major championship wins as a member of the Australian national team. One was near the beginning and the other near the end of my career, 9 years apart. These wins were quite different in many respects; however, they were both world-beating performances. The one thing in common for both, and in general for all archers who achieve at the highest level, was a lot of training. Most of this training was actual shooting—up to 7 hr or 300 arrows a day—but training also included up to an hour of running (or similar aerobic activity) as well as strength sessions 3 to 4 days a week. The competitive climate and culture of the sport at the time of each win, however, were quite different and led to quite different preparations and "tapers."

SIMON FAIRWEATHER

Australian archer Simon Fairweather takes aim during the semifinal of the men's individual event at the Olympic Games in Sydney on September 20, 2000. Fairweather won the semi and then went on to win the gold medal by beating American Victor Wunderle in the final.

© AP Photo/Greg Baker

Simon Fairweather won the individual archery gold medal at the World Championships in Poland in 1991 at the age of 21. Nine years later, Simon won individual gold again, on home soil, at the Sydney 2000 Olympic Games, equaling the Olympic record for an 18-arrow match in the first round. Simon has participated in five consecutive Summer Olympic Games: Seoul 1988, Barcelona 1992, Atlanta 1996, Sydney 2000, and Athens 2004. Simon was declared young Australian of the year in 1991 and 1992.

1991 World Archery Championships

My first major win was the World Archery Championships (WAC) in 1991. I was 21, had been shooting (Olympic-style archery) since I was 17, and had no coach. The two major events I had been to before this were the 1988 Olympics and the 1989 WAC. I had achieved reasonable results, 16th at the Olympics and 20th at the WAC, and these made me determined to do better. I practiced a lot. I adjusted my shooting to match the format of the event. Distances changed every nine arrows back then, so I trained to change every six, so I was accustomed to the changes. I lifted weights and ran: lifted more than ran. I shot a heavy weight bow, at least relative to what everyone shot at that time, and I believed in what I was doing.

The team climate at that time was very amateur, much different than now. Most of the competitors had paid their own way, and the experience was something of a holiday as well as a competition. We were all there very much for archery and our love of the sport. We sat in cafes and relaxed and spent as much time talking about archery as doing it. It was an engaging, supportive, and encouraging culture, and being at the beginning of my career, I felt positive about a long and successful future.

In the lead-up to the 1991 WAC, we traveled to Europe with plenty of time to get over any jet lag. We shot a warm-up tournament in Italy and then moved to Poland, where the event would be held. We arrived about a week before the championships was due to begin. With time on our hands and not a lot to do, the temptation would have been to practice all day; however, we could only access the practice facilities when the organizing committee allowed us to and when we could get transport there. In addition, there is limited ability to shoot a lot of arrows when you are sharing a practice range with other teams that are fine-tuning their techniques and equipment. Tapering the amount of shooting was (and still is) just part of being at the event.

From the time I left Australia I ceased my weight training and running. I believed from magazine articles I had read that I would not lose fitness in the 2 weeks and that it was best to rest and conserve energy. Accommodations in Poland were very basic; it was hard to get good food, and we had no support in relation to our logistics. But I was young and excited to be there, and none of the difficulties affected my physical or mental preparation. The result was that I was rested, relaxed, and confident; my performances improved as the competition went on; and in the end I emerged as world champion.

Seizing the Moments

As a result of the decreased time spent shooting, lifting, or running and increased time spent waiting around for the first day of competition, you get to think a lot more about shooting than to actually do it. This can be good or bad, and how you handle it is vitally important. In 1991 I gained both calmness and confidence from discussing archery techniques and equipment with the other archers. I focused on my strengths in both how my shooting was feeling and the physical fitness I knew I had. Throughout my career, my physical preparation was always important to me and I believed that it gave me an advantage over my competitors. Choosing to think and act constructively during the down time turned what could have been a negative experience into an opportunity to hone my mental edge.

2000 Olympics

Nine years later, my other major win was at the 2000 Olympic Games, for which the preparation was quite different. At this event I had a coach. He was a well-respected coach from Korea (which is the powerhouse country in archery), and the entire Australian national team had worked with him for 3 years leading up to this competition.

The quantity of my training program in the lead-up to 2000 was similar to 1991, although there were some differences in the specifics of the training. By this time, I had increased my running and cut back on the weightlifting, believing that this combination was better for my shooting style. I was confident I had superior fitness compared with my competitors. The competition format had changed from shooting individually, with best total score winning, to match-play, which involved fewer arrows and therefore less margin for error. In training we did a lot of competition preparation that was specific to the demands of the match-play and the competition environment. The national coach had spent the previous 3 years rebuilding our shooting techniques to something he believed was more effective, which at times I had struggled with. I found it hard changing a technique that I had essentially "self-discovered" over many years.

At this time there was more pressure in both the format of the competition and the Games themselves than I had experienced previously. Mental preparation was more important than ever. This was my fourth Olympics, and both my teammates and I had a feeling that time was running out and this was our last chance. The challenge was not to just sit around and stew but to keep positive and occupied. This time rather than completely dropping my fitness work, I reduced my running volume but continued to do some easy running to help relax and burn off excess energy. The decrease in training opportunity and time was the same, but the environment was not nearly as relaxed as that before WAC in 1991. The main difference between the two experiences was that in 1991 I had little if any pressure and I gained reassurance and confidence largely from the camaraderie of my fellow archers. In 2000, there was a lot more pressure, but I received support from my coach and other professionals supporting the team. Both times, however, I was able to optimize mental and physical energy at the right time.

At the beginning of the final week before the archery competition at the 2000 Olympics, we went to a small local archery club to get in a little easy shooting. I was focusing on some technical points of my shooting, because something seemed amiss. We made a little change and for some reason it triggered vivid memories from the 1991 championships. I was suddenly filled with hope, confidence, and positive feelings. I rode that feeling for the rest of the week. I focused on just the positives and the process, taking it one step at a time and never getting too far ahead of myself. It was the perfect culmination of all my years of shooting and competition experience. On top of this I had made the decision to retire after this event. The result of this was that I kind of "let go" of chasing success, which ironically freed me up just to focus on making the most out of this competition, put in a last big effort to do things well and to keep my head in the right space. The lead-up, environment, and even culture of the sport were different than during previous competitions, but my mental and physical preparedness was very similar and led to the same winning result.

State of the Art or State of Mind?

More and more coaches look for science and technology for improving their archers' performance. Lots of money can be spent on sport science, including studying things like tapering, but I'm really not sure that it is money well spent in a sport like archery. There were many occasions in my career where we tested things that were thought to provide advantages. In general all it did was interrupt training and mess with my confidence. I acknowledge that many sports gain a lot from sport science, but I think that in high-skill, low-physicality sports like shooting and archery, the real strength for performance comes from confidence and routine. I believe that an archer feeling really confident and comfortable, with a positive outlook, will beat a technically superior rival who is struggling with confidence and full of doubt. It certainly is much cheaper to work on confidence than equip a training center with state-of-the-art technology. You only have to look at the results of many of the poor European countries: They often have substandard equipment and just basic infrastructure yet they are a force to be reckoned with. State-of-the-art equipment obviously doesn't make much of a difference. Maybe it only adds to the athletes' drive to succeed.

Another thing to keep in mind is that lack of planning by coaches and athletes, and mismanagement by team officials, can affect team members badly. Even simple things like having enough water for the athletes—not just after you've settled into a routine but during travel and certainly as soon as possible once you've arrived—can affect their mental balance. Confidence and peace of mind are often fragile things and it's really important to protect them.

Tapering for Archery

I believe that confidence, mental energy, and preparedness are the most important parts of the "taper" for an archer. The best way to achieve this is to take a very measured, calm approach to the tasks of the day, ensuring that there are enough activities in the day to occupy the mind of the athlete without causing fatigue. Obviously this is one of the challenges for the coach and manager, and the competition infrastructure may not make this easy. So coaches should look for ways to improve their athletes' outlook, such as these ideas:

- Be sure athletes develop good general fitness in the lead-up to the event.
- Try to make traveling as easy and comfortable as possible.
- Assign athletes some easy exercise upon arrival to use some of the energy, adrenaline, and tension that can build up.
- Set aside time for relaxation sessions. By *relaxation* I mean a technique such as progressive muscle relaxation, not a trip to the pub!
- Provide massage, if funding permits. Soft tissue therapy is not just a good physical treatment but a good mental one too. It takes up more of the day with something that feels and *seems* good for the competition preparation.

- Avoid getting stuck waiting around. Avoiding needless delays, as much as possible, will lessen athletes' anxiety and associated conflicts between team members.
- Provide strategies for coping with the unexpected and ensuring minimal stress.

I don't believe that tapering for archery is as advanced or scientific as it is for many sports. If preparation has been thorough, fitness, technique, and competition-specific practice should set the athlete up well for the major event. The athletes should keep the "feel" of the skill or technique with some shooting each day but significantly reduce the time spent standing outside in the elements, thereby conserving energy and strength for the competition. Physical preparation is fairly simple and based on common sense: Aerobic and strength training can be dramatically reduced 2 weeks out from the event. I have never heard an archer say that his competition was affected because he didn't get his taper right. Mental preparation is the key and there are many factors in optimizing this, but the perfect mental taper is elusive . . . and not just in archery!

Table 11.1 shows a daily program that will help build and maintain confidence, mental energy, and preparedness in archers in the week leading up to a major competition.

Table 11.1 Daily Program for the Week Leading Into an Archery Competition

	Monday	Tuesday	Wednesday	Thursday	Friday
Morning	Arrive. Check in to lodgings. Jog or walk.	Have an easy morning, not too early. Spend 3 hr shooting.	Start earlier, similar to a competition day. Spend 2 hr shooting.	Start as yesterday. Spend 1 to 1 1/2 hr shooting.	Start as yesterday. *Official practice day;* therefore not too much shooting.
LUNCH					
Afternoon	Spend 2 hr training. Nothing specific, just easy shooting, allowing time to catch up with old friends.	Check shooting equipment. Refletch arrows if needed. Jog.	If possible, spend another 2 hr shooting, identifying key points to remember for good shooting (different for each individual).	If possible, spend another 1 to 1 1/2 hr shooting, identifying key points to remember for good shooting (different for each individual)	Jog—not too far, just to loosen up and take up time. Check and repair any equipment. Go to cafe for coffee or tea.
DINNER					
Evening	Go to cafe for tea or coffee and then have an early night.	Have a massage and practice relaxation.*	Have a massage and practice relaxation.*	Have a massage and practice relaxation.*	Don't have a massage today, because it often affects shooting feel. Practice relaxation.

*A therapist has time to see only so many people per day, so not all the team may get a massage each day. Some may not like it daily anyway.

　Applying Simon Fairweather's Expert Advice

In this section by Simon Fairweather, readers learn how a world and Olympic champion approaches the final days before a major archery event. Although the notion of tapering for competition is not embraced as such in this sport, there are hints of a true tapering approach in Simon's account. Indeed, he mentions a significant reduction in his fitness training and arrow shooting, optimization of his psychological status, and individualization of the optimal peaking strategies in the final days before a major event. Key concepts of this section are adjusting the training to the specific requirements of the competition; incorporating travel and limited access to training facilities at the competition venue into the tapering program; ensuring that logistical issues do not interfere with an athlete's preparation; and focusing on strengths and self-awareness of the advantages an athlete has over her competitors to achieve the confidence and mental energy needed for optimal performance.

José María Olazabal

Fine-Tuning for Peak Performance in Golf

Golf is an activity that allows the player, even at the professional level, to maintain a high level of competitiveness and achieve a certain longevity. I mention this because the long sporting career of a golfer is often characterized by many changes in the competition calendar, and these players go through many trends regarding physical, mental, and technical preparation.

I have achieved three individual wins that I especially value and consider particularly important, because of either the competition in question or the circumstances surrounding it. In terms of team victories, I refer to two: the Ryder Cup of 1987 and that of 2006. My special individual wins are the 1990 World Series and the Masters of 1994 and 1999. I describe these in chronological order.

JOSÉ MARÍA OLAZABAL

Golfer José María Olazabal of Spain holds the trophy after winning the 2000 International Open at Belfry, UK. Olazabal won by three shots, the second time he has won the event.

© AP Photo/David Jones/PA

José María Olazabal is one of the world's most successful professional golfers of the past 20 years. He was competing at the national level by age 7. At 19, he went professional. His greatest achievements include two victories in the Augusta Masters, in 1994 and 1999, and four victories in the Ryder Cup as a member of the European team, in 1987, 1989, 1997, and 2006. Throughout his professional career, "Txema" Olazabal has won 22 tournaments in the European tour, 6 tournaments in the American PGA tour, and 2 professional tournaments in Japan. In the amateur ranks, he is the only golfer in history to have won the British Boys, Youth, and Amateur Championships, and his amateur record is officially recognized as the best in the history of the sport.

Ryder Cup 1987

It was the first time in the 60 years of history of this competition that the American team was defeated on their home soil. The cup was played in Columbus, Ohio, Jack Nicklaus' "fief," and it represented for me the discovery of a competition I became hooked on and I am thrilled about.

My debut in the circuit in 1986 had been a walk in the park. That year I was winner of the national circuit ranking and two tournaments and was second in the Order of Merit right behind "The Myth" (Seve Ballesteros). So I could not understand why I was not winning in 1987, but it turned out to be a gray season without individual victories in the circuit.

Despite that, Tony Jacklyn, the European Ryder Cup captain, had confidence in me, selected me for the team, and made me team up with Seve in all four double matches. But I arrived to the competition with my self-esteem below zero, to the point that the Irish pro golfer Des Smyth walked up to me the week before at the Lancôme Trophy in Paris and said, "José, you are not hitting it right. I see you fail a lot of shots, and I understand that you get upset, but damn it! You hit quite a few good ones too; try to enjoy them all in the same measure." I did not feel the impact of those words at that very moment, but that idea evolved over time, slowly but surely.

At the Ryder Cup, it could not have happened in any other way; I unconsciously let myself get carried by Seve and the rest of team, and it really worked! It was the first time that the "mind factor" led my performance, rather than the "testosterone factor," and I didn't realize it at the time!

World Series 1990

My first individual victory on U.S. soil was at the legendary course of Firestone, in Akron, Ohio. Only four players were under par, but with a total score of 18 under par, I was 12 shots clear of runner-up. My record of 61 shots still stands, despite technological advances in golfing gear. The course still has the reputation of being brutal.

When I arrived there, I was not impressed with my game and I was not enjoying it. The week before I had had a decent role in Denver, but nothing to be crazy happy about. But it happened. I still don't know why, but I entered "the zone" from the very first shot, like being transported inside a bubble, without external influences, with total inner peace and stillness. Everything seemed to happen in slow motion. I started the tournament: birdie–eagle–birdie–birdie. I finished the first round with a 62 (9 below par), and everyone treated me like I was from another planet. The week continued that way and I am unable to capture why it was so. I imagine that psychologists have theories about it, but I am no expert.

Masters 1994

From March 1992 until February 1994 I played terribly. Of course, I never came close to winning. In February 1994 in Andalusia, I finished second despite not playing particularly well, and I started to think that perhaps I could achieve a decent result without playing brilliantly. The following week in Alicante I played better. I was in the last group on Sunday, but I was three shots behind Paul McGinley with two holes left to play. He collapsed and we tied after the final hole. The tie-break did not escape me, and I beat him.

After 1 week of rest I was off to the United States, where I finished eighth and second in the tournaments immediately before the Masters. Nevertheless, I maintained a low mental profile, without overthinking the competition. It worked well, and I believe that in the final moments my runner-up position in the Masters 1991 behind Ian Woosnam gave me extra strength and leadership over Tom Lehman to defeat him. Once again, my success came more from my head than anything else—not in an induced way, but instinctively.

Masters 1999

Low times again. I had been happy to emerge after my injury in 1996, which kept me away from competition for 18 months, and I had done little things here and there, but things were not rolling.

During the champions dinner at The Masters, Gary Player (three-time winner of the Augusta Masters) came up to me and said, "How are you José, how is your game?" and I replied, "Don't bet any money on me Gary, I don't stand a chance." He became angry and delivered a solid speech about how he, at the age of 70, could still hold karate positions for 5 min, with all that tension on his thighs, and how it was incredible that someone like me could have such negative ideas, so little self-esteem, and so little faith in himself. Then he added, "Now get out there and win it." And the truth is that I did: I beat Greg Norman to win The Masters.

Later I found out that my entourage had asked Gary Player to deliver that speech to me. And it certainly worked—the head once again.

AT A GLANCE   **Applying the Expert Advice of José María Olazabal**

In this section by golf great José María Olazabal, the reader can learn about the power of self-esteem and self-belief in a psychologically demanding precision sport like golf. Unlike other contributors to part III of this book, but in agreement with some others, Txema Olazabal does not mention tapering his training in the lead-up to his major golfing achievements but rather fine-tuning his mental preparation for the competition. This latter concept is in full agreement with the notions about mental readiness expressed by archer Simon Fairweather in the previous section. Another interesting aspect of this contribution is the idea that individual sport athletes can become "mighty locomotives" for a team event such as the Ryder Cup, when they all are on the same wavelength and pulling in the same direction. The key idea from this section is the paramount role of "the mind factor" in a sport like golf, in which hitting the ball right may depend more on achieving the right mindset at the right time than on any other training aspect.

Ryder Cup 2006

It was my seventh participation in this tournament, which was in Ireland this time. I understood for the first time the meaning of collective catharsis. None of us on the team were ourselves. We were 12 souls, plus that of the captain, and 13 souls pulling in the same direction represent a mighty locomotive. We won, we were unstoppable, and it was almost by collective hysteria. The motives were, for example, Darren Clarke's recent widowhood, the passion expressed by the Irish crowds... and I don't really know what else, but the reality is that it happened.

We all prepare physically to play, but we have a long season, the major tournaments are scattered through the year, and you should be in good form from mid-January to the end of the year, unlike other sports in which the major event takes place once a year and one prepares "in crescendo" to reach the summit at the right date. Therefore, what needs perfect timing in golf is one's mind, and that timing is not easy to achieve.

chapter 12

Tapering for Team Sports

As we have seen in chapter 8, there are some unique aspects of tapering and peaking in team sports that coaches and athletes should always keep in mind. Team sport performance requires a perfect balance between physical, physiological, technical, tactical, and psychological factors. Elite team sport players must excel in their speed, acceleration, power, endurance, and agility, and an effective taper must optimize all of these attributes for every player. In addition, team dynamics must be developed in such a way that every piece of the puzzle falls into place, every member of the team is singing the same tune. In this section, an outstanding group of coaches from around the globe disclose their approach to exceptional world-class team sport performances.

Ric Charlesworth

Achieving Gold in Women's Field Hockey

Our strategy before Sydney 2000 was to peak physically, mentally, tactically, technically, and in terms of our cooperation and team dynamics. That is, we worked to be at our peak in every aspect of the game during the tournament, with a special emphasis on the necessity to be in the best position to play at the end of the tournament. In a team event with support staff in double figures and 16 players, this is not easily controlled or supervised, and it requires real vigilance by all team members to ensure that some are not left behind and all are included in the plans and are on the "same page" as teammates. Equally, all must be aware of each other's performance strengths and weaknesses.

Every coach aims to optimize performance at the major events and develops the tapering and peaking strategy to achieve this. During the years before the Sydney 2000 Olympics, we played out various scenarios to ensure we had our group ready for the last 2 weeks in September. Our final, if we made it, would be September 29. Starting on the 17th we would be required to play 8 matches in 13 days . . . the field hockey tournament at the Olympics is an endurance event as much as anything else!

6 Months Out

Our training regime for the 6 months prior to the games was structured around four game-simulating sessions per week. At these sessions, intensity was tracked through heart rate monitors and lactate measurements to ensure we matched elite game levels of physical

RIC CHARLESWORTH

Australia's Jenny Morris (left) and Allison Peek celebrate after scoring during their gold medal field hockey game in the Sydney 2000 Olympics. Australia won 3–1, bolstering the case for coach Ric Charlesworth's carefully planned tapering strategy.

© AP Photo/Eric Gay

Ric Charlesworth was the national coach of the Australian women's field hockey team from 1993 to 2000. The team was ranked number one in the world for eight consecutive years, won the gold medal at the Atlanta 1996 and Sydney 2000 Olympic Games, and won the gold at the World Hockey Cup in 1994 and 1998. During this time, Ric won the Australian team coach of the year award on six occasions. In 2001 Ric was appointed a master coach by the International Hockey Federation, and in 2005 he was selected as Western Australia's greatest-ever coach.

output. We understood what the requirements of the game were because for years we had collected heart rate data from our international matches. We constantly evaluated training intensity in these sessions to ensure we exceeded match requirements.

We used a volume–intensity cycle three times during the specific preparation phase of 23 weeks. This was preceded by a regeneration phase at the end of the previous year's activities, and then through December 1999 and January 2000 there was a general preparation phase to prepare for the intense program ahead. The players were relocated from their various home bases to a single, central base at the beginning of February 2000 and the general preparation phase continued into March.

The specific preparation program then began, and as already indicated we used a volume–intensity cycle three times during that program. The final taper duration was longer than that of the two previous phases, because the focus was more on ensuring that the players carried limited structural injuries and fatigue into the competition. We knew by then that their physical capability under maximum load returned after 12 to 16 days of light work and within 3 to 5 days if one or two hard sessions (game simulations) were held during this time.

Throughout the specific preparation phase we trained intensively for segments of 3 to 4 weeks followed by a competition phase and then rest, and so the players were accustomed to these phases and workloads. Generally, we lightened the loads prior to competition phases and we were well aware of the players' capacity to return to maximum physical capability.

Foundations for the Final Taper

Immediately before the Olympic Games began, our lightest week was 2 weeks before the beginning of the playing phase on September 17. The taper also took into account that the team should physically peak for the final games, and therefore the first week's competition would be part of the load required to ensure this occurred.

Critical to the effectiveness of the physical preparation strategy was the work done over the previous years with heart rate, lactate, and perceived ratings from the players to ensure their ability to self-monitor, self-regulate, and provide feedback. This gave us confidence that we could handle the demands of competition and that the athletes would respond as predicted to physical loads. Just as important as the physical loading and training experience of the group was the confidence that came from the data showing that they could handle the physical requirements of eight games in 13 days. This confidence was as important for their performance as was their physical capability.

Technical peaking was underpinned by sessions that were, as always, match intense and competitive. During the two light weeks immediately before the games, these sessions were of shorter duration but of the same quality, pace, and intensity (September 7, 11 [short game], and 13 [short game], table 12.1). Equally, the players usually felt more comfortable about their technique when they did some training every day. We trained throughout the competition as much for a distraction and time filler as for any specific training aims.

Unique Olympic Team Issues

In the Olympics you can find yourself playing Korea one day and Spain the next. These opponents are very different in style and approach, and the work done over the months and years before must ensure that there are no surprises. Both the physical and tactical preparation must instill total confidence in the players. This ensures there is no panic about what they will face in the weeks ahead. The last days of specific preparation were spent summarizing and revising important tactical knowledge that we had systematically accrued over the previous years and refined in the 6 months of specific preparation in 2000.

Another big issue with the Olympic hockey event is that over the 2 weeks fortunes will wax and wane, and the ability to remain grounded is critical. Too much time to think about the event can be a problem as can the short-term disappointments or elations that occur one day and must be set aside to focus on a new opponent the next day. Results of other matches can similarly distract and disorient individuals, the squad in general, and the staff; thus, it is necessary to meet regularly to discuss the events of the day and monitor formally and informally the feelings and thoughts of the group. This practice should be established over the years leading into the Olympics.

Critical to our frame of mind were some matters that I believe are essential for Olympic team performance. We selected late so that the whole focus was on improvement and every squad member's being ready to play, not just "make the team." Also in 1998 we had Olympic-like experiences in the World Cup in The Netherlands, where we shunned superior accommodation for the village lifestyle, and the Commonwealth Games in Kuala Lumpur, where we had a similar village experience. We knew what to expect. Finally, the family and parents week preparation early in 2000 dealt with all the issues of family, friends, and media that a home Olympics can bring. The expectations of others were appropriately shaped and athletes' expectations were well matched to reality well in advance.

5 Weeks Out

In the 5 weeks before the Olympics we went from as many as 15 sessions a week down to 6 in the week before the games commenced and 7 the week before that (table 12.1). Indeed, the week before the Olympics was the lightest week of the year. However, our past experience gave us confidence that physically the athletes would respond as we had seen previously. During the 14 to 7 days before the Olympics we relocated to Sydney, became familiar with the village, and then after a couple of days moved out of the village to train and stay in the Blue Mountains, away from all of the hype associated with the Olympics. During this phase no hockey was played; however, we had a number of "novelty" sessions and worked on team dynamics, aerobic conditioning, and weights training. One of the novelty-sessions days entailed playing games of "keep off" and relay races in a pool.

We then moved back into the village for the week before the games, where we played a couple of short practice matches, did some sharp speed work on another day, and did some technical preparation prior to a 2-day rest before our first match.

Table 12.1 shows that in week 44 we had our traditional four high-quality intensity sessions on August 21, 23, 24, and 26/27. In week 45 sessions were held on August 28, 30, and 31 (this was eventually cancelled), because some players went on a home leave. The intensity was accordingly winding down.

Table 12.1 Australian Women's Hockey Squad 2000 Olympics Training Program, Specific Preparation III and Taper

Week	Phase		Monday	Tuesday	Wednesday	Thursday	Friday	Saturday	Sunday
44	Specific preparation	Day Dates	21 Aug	22 Aug	23 Aug	24 Aug	25 Aug	26 Aug	27 Aug
		a.m.	Active recovery GK weights	Weights Set plays		Set plays	Rehab weights Set plays	GK weights	Skills high tempo or club semi
		p.m.	Sprints and skills (moderate to high tempo)		Game and RE 20 min	Game			
45	Specific preparation	Dates	28 Aug	29 Aug	30 Aug	31 Aug	1 Sept	2 Sept	3 Sept
		a.m.	Active recovery GK weights	Weights Set plays		Set plays	Rehab weights Set plays	GK weights	Field AE run 30 min
		p.m.	Sprints and skills (moderate to high tempo)		Game and RE 20 min	Game cancelled			
46	Taper	Dates	4 Sept	5 Sept	6 Sept	7 Sept	8 Sept	9 Sept	10 Sept
		a.m.	Rest	Enter village	Training (WU pitch)	Team weights			
		p.m.	Skills (comp pitch) or light jog and stretch	Speed and skills L-M tempo (comp pitch)		Skills high tempo (comp pitch)		AE 40 min with surges 10 s every 4 min	Skills (comp pitch) L-M tempo

47 Taper

Dates	11 Sept	12 Sept	13 Sept	14 Sept	15 Sept	16 Sept
a.m.	GK and rehab weights	Skills (comp pitch)	Skills (Ryde) set plays		AE and up tempo skills 40 min (9:30am) Ryde	Light jog and stretch
p.m.	**WU game vs. Germany** (comp pitch)	Training (WU pitch) sprints	Training (WU pitch) **game vs. Holland**	Skills (comp pitch) L-M tempo	**Opening ceremony**	**Game vs. GB**

48 Competition

Dates	18 Sept	19 Sept	20 Sept	21 Sept	22 Sept	23 Sept	24 Sept
a.m.	Active recovery		Active recovery? (depending on schedule)	Active recovery		Active recovery	
p.m.		**Game vs. Spain**	**Game vs. Argentina**		**Game vs. Korea**		**Game 5**

49 Competition

Dates	25 Sept	26 Sept	27 Sept	28 Sept	29 Sept	30 Sept	31 Sept
a.m.		Active recovery		Active recovery		Active recovery	
p.m.	**Game 6**		**Game 7**		**Final**		

AE = aerobic endurance efforts; comp pitch = location is competition pitch; GK weights = goal keepers' weights; L-M = low to medium intensity; RE = repeated explosive efforts; Ryde = training site in Sydney; WU = warm-up. Bold type denotes key high-intensity training sessions and preparatory matches.

We entered the village on the September 5. After a couple of days settling in and training we left the village for 3 days in the mountains with no hockey. Our nonhockey vacation of 3 days was designed to escape the Olympic environment. We had tested this approach in 1996 and we knew the players appreciated a respite from village life. We returned to the village for more light work, two short (50 min) but high-tempo practice matches on September 11 and 13, and a final sprint session on September 12. We were then ready to play.

Our performance in the tournament indicated that we were probably at our best in the second week. The first half of our first match of the tournament was quite sparkling in its quality, although during the first week of the Olympics we were not at our best-ever form.

Our strategy to perform in Sydney encapsulated a physical taper that was not experimental but had been tested, yet still relied heavily on the players to self-monitor their diet, hydration, sleep, and general health. Thorough preparation for tactical issues that might arise and the development of flexibility over the previous years enabled us to anticipate and react to any eventuality. In the end a solid routine, discussed and agreed on well before the event, underpinned this preparation to ensure we remained on course for the 2 weeks of competition.

AT A GLANCE   **Applying Ric Charlesworth's Expert Advice**

In this section by Ric Charlesworth the reader can learn about the strategies for collective peaking of a group of elite athletes to win Olympic gold. Team sport peaking involves not only optimizing physical, technical, tactical, and mental skill but also enhancing inter-athlete cooperation and group dynamics. Key ideas of this section include matching or exceeding the physical output requirements of competition during training; taking the first week of competition as part of the loading phase to actually peak for the final games of the tournament; training throughout the competition to maintain players' technique proficiency, achieve specific training aims, and also fill time; performing a taper that had been tested and proven to be successful in prior events; and being flexible to anticipate and react to eventualities.

Derik Coetzee, Yusuf Hassan, and Clint Readhead

Winning the World Cup in Rugby

When Jake White was appointed as the South African rugby coach in 2004, he informed his management team and players that there was no reason they could not win the 2007 Rugby World Cup (RWC). From that point onward, players and the management team systematically synchronized and controlled all inputs to fulfill the coach's prophecy.

Designing a training and tapering program according to scientifically based periodization principles was our point of departure. We also relied heavily on practical experience. Tapering is a training technique designed to reverse the training-induced fatigue that occurs during intense training, without losing any of the acquired training adaptations. Tapering is acknowledged as the final phase of training within a time frame (weeks to days) prior to the actual competition and involves a reduction in training load, by manipulating training variables (intensity, frequency, and duration). Yet we also understood that we needed to apply throughout the season the rest and recovery needs that are key to the taper itself. Thus an important part of achieving our goal was to apply some of the principles of tapering to our training year-round.

DERIK COETZEE, YUSUF HASSAN, AND CLINT READHEAD

Bryan Habana and coach Jake White of the South African Springboks hold Rugby World Cup after the 2007 final match between England and South Africa in Saint Denis, France. South Africa bested England by 15–6.

© AP photo/Francois Mori

Derik Coetzee is the head conditioning coach of the South African national rugby team, the Springboks. Derik has been coaching elite rugby players for more than 15 years and has been involved with the Springboks since 2004. Yusuf Hassan is the medical doctor of the Springboks. Yusuf has provided medical services to elite rugby teams for nearly a decade and has been the Springboks' medical doctor since 2004. Clint Readhead is the medical manager for South African rugby. Clint has been a sport physiotherapist with rugby teams since 1995, and he has worked with the Springboks since 2003.

Avoiding Injury and Burnout Through Periodization

Jake White stressed the need to have players at test level who are physically superior to the opposition in every way. He also understood that having the Springboks at peak fitness levels for the duration of the entire international season was not realistic. In accordance we followed a well-planned strategy based on the *periodization* method of training. The goal was to ensure that the Springboks' physical and mental abilities were optimal at the onset of the RWC and that those abilities did not deteriorate during the tournament. This approach required that adequate rest and recovery be built into the training throughout the year—not just the taper—to avoid serious injury and burnout. To do this, we took medical, physiological, technical, tactical, psychological, and their sociological variables into account. Two such factors that are closely associated with tapering and that we continued to watch closely during the taper itself were controlling fatigue throughout the year and minimizing injuries.

Controlling Fatigue Throughout the Year

Players were required to complete a weekly subjective rating of the training intensity and their state of fatigue (table 12.2). Individual ratings were evaluated in view of the actual training loads, and training loads were subsequently manipulated and recovery strategies implemented accordingly.

A special emphasis was placed on recovery strategies during the competition. These strategies included individualized guidelines for optimal hydration, nutrition, nutritional supplement intake, specific recovery strategies, and appropriate rest and sleep patterns at different times of the week and before, during, and after each game (table 12.3).

Minimizing Injuries

A follow-up of Springboks rugby players over two seasons showed that only 12% of the players were able to complete an 11-month rugby season without a significant

Table 12.2   Weekly Assessment of Training Intensity and Fatigue Index

Player's rating of the intensity of the previous week's training (1-10)	Please indicate your level of perceived intensity of the previous week's training in your diary every Friday		
	Reassessment		
	Date 1	Date 2	Date 3
0 = rest			
1 = really easy			
2 = easy			
3 = moderate			
4 = sort of hard			
5 = hard			
6 = harder			
7 = very hard			
8 = he tried to kill me			
9 = oh-oh			
10 = death			
Player's subjective rating of present state of fatigue	**Please indicate your level of perceived fatigue of the previous week in your diary every Friday**		
1 = very very slight			
2 = very slight			
3 = slight			
4 = mild			
5 = moderate			
6 = moderate to severe			
7 = severe			
8 = very severe			
9 = very very severe			
10 = totally exhausted			

acute or overuse injury. Indeed, 52% of the players sustained overuse injuries, some of which required surgery. Of the 48% of players who did not suffer any overuse injury, 73% had acute injuries requiring rest and rehabilitation, which lasted for weeks in many cases. Only 27% of the group without overuse injuries played the full duration of the 11-month rugby cycle without significant acute or overuse injuries (figure 12.1). Because of these observations, it was concluded that ongoing screening for biomechanical intrinsic risk factors must continue so we could identify athletes who were most vulnerable to injury under the training and competition regimes that were being implemented. Most of the contracted Springboks had been training and playing during the preceding 2 years, so because of the way contracting is done South Africa, it was difficult to control the total amount of rugby a contracted player was exposed to. Thus, we predicted an increase in risk of injury and burnout.

Table 12.3 Guidelines for Recovery During the Competition

Beginning of the week	Measure body weight at breakfast Monitor fluid Maintain flexibility Maintain mobility Have massage Maintain high protein, high carbohydrate intake Follow individual plan for nutrition and supplementation
Middle of the week (main training session)	Maintain high protein intake Maintain hydration Rest Stay off feet Sleep at least 8 hr
End of the week	Engage in light flexibility work Have massage Maintain high fluid intake Maintain high carbohydrate intake Rest Stay off feet Sleep at least 8 hr
Pregame and game	Maintain high fluid intake Maintain high carbohydrate intake Maintain fluid intake during game Ice jackets compulsory half time New jersey Bench thermo suits Rest Stay off feet Sleep at least 8 hr
Postgame	Maintain high fluid intake Maintain high carbohydrate intake Maintain light locomotion Engage in light flexibility work Engage in cold immersion Avoid alcohol

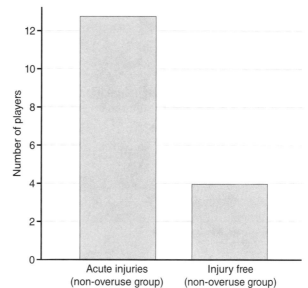

Figure 12.1 Proportion of overuse-injury-free Springboks players who suffered other types of acute injuries compared with those who suffered none.

Adapted courtesy of Drs. Yusuf Hassan and Derik Coetzee.

Because of our obligations of not interfering with the scheduled *Super 14 club competition*, we made it a priority to identify and manage players who had injuries or needed rest. Players were given time off based on their medical profiles, physical strengths and weaknesses, as well as their rugby exposure times. The affected players were then introduced (according to their fitness profile images) to the various stages of the periodized protocol described in the following sections. Also certain players who presented with "specific risk of injury" profiles were strictly managed in terms of their training load exposure. These strategies had a huge effect in that we did not see the normal overuse injuries with these players when they reported to the Springboks camp in May 2007 for the start of the test season.

Rest and Recovery in Pretaper Training and Play

To meet the optimal required standards, a core group of players were identified and trained from 2004 to 2007 to make sure that appropriate physiological adaptations (being able to run faster, jump higher, lift heavier weights, increase muscle mass, and simultaneously reduce body fat) would be achieved. But at the same time we had to avoid overtraining. Following are the pretaper strategies we adopted.

After the Tri Nations in 2006, the contracted Springboks were rested for 2 weeks and this was followed by an intensive conditioning program for 7 weeks. At the end of the 7 weeks the players returned to their Super 14 franchises and rested for 3 weeks (the December rest phase). All players resumed duties (in terms of Super 14 training obligations) in the first week of January 2007.

The players were now well rested and the acquired adaptations were maintained. In planning we had reasoned that playing time would allow players to develop quality playing form. With the help of the Super 14 medical and conditioning teams, we tracked and recorded all the players' training and match times as well as injuries they sustained and how they were managed. While the Super 14 was on track, the national medical and conditioning staff met every month with the national coaches and planned the training program that the returning players would undergo. The program detail was such that we could break down the volume and intensity of each session from our assembly date on May 16, 2007, right up to and including the captain's run the day before the 2007 RWC final.

We assembled on May 16, 2007, in Bloemfontein and played the first test 10 days later. We played with our World Cup squad for the three Vodacom Incoming tests (England twice and Samoa) and two home Tri-Nations Tests. Following these five matches, 20 players (representative of the core World Cup squad) were identified and withdrawn according to their physical and medical attributes to refrain from competing in the second leg of the tri-nation tournament. The intentions were set to evaluate and continue with conditioning programs in preparation for the World Cup.

Following this high-intensity, high-volume training segment, the players returned home for 1 week (rest and psychological purposes). On July 21, 2007, the Springboks RWC squad was announced and the team entered the last phase of preparation for the RWC. The last phase of preparation lasted 5 weeks and consisted of a 2-week high-intensity training strategy followed by the warm-up test (game) against Namibia and two games abroad (Connacht in Ireland and then a test match against Scotland). Between these test matches the players were subjected to moderate loads to accommodate for the effects of traveling and training (fatigue) while maintaining the adaptations previously acquired.

Tapering Strategies for the 2007 RWC

On our return the squad was sent home for 4 days to recover physically and psychologically before we assembled on August 31 in Johannesburg to depart for the RWC, where the first match was to be played within the next 8 days. During this time the emphasis was on technical and tactical training to minimize the risk of injury.

After the first match, we adopted an exponential strategic taper technique with a fast time constant of decay after the pool games. The training load was reduced significantly for the quarter finals against Fiji, the semifinals against Argentina, and last the final against

Captain's Run

The captain's run is a training session that is performed the day before a test match at the venue where the test match is to be played and as close as possible to kickoff time. The intensity of the training bout is usually between 60% and 80% of maximum. It is a structured exercise session where the captain takes his team through the game plan and they execute the various planned moves for the test match. The captain's run gives players a feel of the potential weather conditions that may prevail at that time on test match day and allows the players to begin mentally focusing on the job at hand and what they have to do to win the next day's test match.

England. The primary aim of the taper technique between games was to minimize the accumulative fatigue index buildup when competing in consecutive pool games rather than to attain any physiological adaptations or fitness gains. The maintenance of training intensity and the exercise-to-rest ratio were carefully monitored by means of the players' subjective ratings of perceived intensity and fatigue (table 12.2).

We believe that the team was successful because the required peak fitness level of players (4 years of preparation) was maintained for the entire duration of the tournament (tapering strategy included). This was accomplished by implementing a sound, realistic, well-planned periodization strategy incorporating all medical, physiological, technical, tactical, psychological, and sociological inputs. Surely, practical experience and a bit of luck also could have made a significant contribution.

Applying the Expert Advice of Derik Coetzee, Yusuf Hassan, and Clint Readhead

AT A GLANCE

In this section by Derik Coetzee, Yusuf Hassan, and Clint Readhead, readers can appreciate the importance of establishing a collective goal such as winning the Rugby World Cup and implementing a coordinated strategy involving not just the players but all the support staff in an effort to achieve that goal. The Springboks benefited from training programs that incorporated both scientifically based periodization principles and practical experience. Key ideas from this section include the application of some of the rest and recovery principles of tapering to the year-round training; quantifying training loads and players' subjective perceptions of fatigue; finding the optimal balance between training and recovery; and minimizing the incidence of injury throughout the competitive season, in a coordinated effort between players' club staff and national team staff.

Greg McFadden

Peaking for World Championships Silver in Women's Water Polo

This section discusses what we have tried to achieve with the Australian women's water polo program from January 2005 to December 2007. The main focus over these 3 years was to create a stronger depth of international players and return Australia to being a medal contender in all major competitions. The ultimate goal was to win gold in Beijing 2008, and although we fell short of that, it wasn't by much: We came home with a bronze. With further refinements in regular season and peaking phase training, we're aiming again for gold next time around!

GREG MCFADDEN

Greg McFadden has coached water polo teams for almost 20 years. Greg has been the national coach of the Australian women's water polo team since 2005. Under his direction the team is consistently ranked among the world's top 3, winning medals at all major competitions: bronze at World League 2005, gold at World Cup 2006, silver at World Championships, World League 2007, and bronze at the Beijing 2008 Olympics. In 2006 Greg and his team won the Australian Institute of Sport coach of the year and team awards, respectively. Both were finalists in these awards in 2007.

Australian coach Greg McFadden congratulates his players after their 12-11 win over Hungary during the women's water polo bronze medal match at the 2008 Beijing Olympic Games.

© Jason O'Briena/action images/ICON SMI

We will look at our preparation for the 2007 World Championships in Melbourne, in which we made it to the final and won the silver medal. Ultimately we lost by 6 to 5 to the United States in a game that could have gone either way. Australia had three good goal-scoring opportunities in the last 40 s, only to see two of them repelled by the U.S. goalkeeper and the final one called for an offensive turnover. The United States has been the most consistent team in the world over the last 7 years, having won medals at every major championship except the 2006 World Cup.

Tapering and Peaking for Major Events

In team sports such as water polo, you have to consider the sacrifices you are willing to make in the lead-up to tournaments along the way to your major goal and competition. You also need to consider that the team consists of 13 players in which only 6 field players and a goalkeeper are in the water at one time. Generally, the field players are interchanged regularly; however, during the competition not all players receive equal playing time. This leads some players to detrain if they don't receive extra work during the lead-up to tournaments. These players need to maintain their fitness and be at their peak for the major competition.

The reverse has to be done for the players who receive a majority of pool time during these games, and so recovery or lighter training sessions are required for them. The ultimate goal is to get all 13 players as close as possible to their physical peak for the major competition.

During the lead-up to the major competition, athletes should have 2 to 3 days off training with the exception of a light strength training session incorporated into this break, so that when the athletes return to training or the major competition they experience minimal muscle soreness. This break happens 5 to 10 days before the start of the major competition and is just as important mentally as physically. We want players to be refreshed and to forget about what lies ahead. The timing and length of this break depend on what has been done in the prior 6 to 10 weeks.

Key Issues During Tapering and Peaking

The areas that are critical to successful tapering and peaking are the intensity and duration of training; technical skill work and tactical teamwork; recovery; nutrition and strength training; and psychology and group dynamics.

Intensity and Duration of Training

Prior to major competitions in water polo, teams participate either in lead-up tournaments or training camps with opposing countries. This allows us to play games that are very close to the intensity of the major competition we are preparing for. But winning is important, too, for psychological reasons.

We also need to be sure, as noted, that the best players do not become too fatigued and the remainder of the players don't lose their fitness (detrain). During training camps this is a lot easier to control because you can make sure that all players have close to the same pool time. However, in the lead-up tournaments this is a lot more difficult. You want to be competitive, and trying to win becomes more important in your preparation because going into a major competition with a poor win strike rate can lower the players' confidence. Also, the referees' expectations of your team will be lower if you have been showing poor results.

To help us track pool time, we use a daily athlete monitoring sheet and computer program, which the players fill out online. If you have had a good and uninterrupted training preparation in which you have built a very good fitness base prior to these lead-up tournaments and training camps, maintaining fitness levels during this time can be as simple as doing one or two fitness (swim or game simulation) sessions per week.

As the major competition becomes closer, the duration of the training sessions is decreased but the intensity of the sessions remains high. For example, instead of playing four quarters of water polo you would cut this down to two quarters, however insisting that the sessions are played at 100% in terms of quality and intensity.

Technical Skill Work and Tactical Teamwork

During the preparatory training phase, the team undergoes seven 2- to 3-hr training sessions a week involving basic skill work, individual position skill work, and tactical teamwork. Because our sport is water based, we can play full or modified games at full intensity. This allows for about 75% of the tactical teamwork to be done under full game conditions. But during the peaking competition phase we also have to fit in our fitness and strength training, so we reduce the individual skill and tactical team sessions to three to four times a week and only 60 to 90 min a session. These sessions become more specific, and we focus on the specific component of the game that we need to improve (e.g., 6-on-6 defense, 6-on-5 attack).

During the major competitions such as the World Championships or Olympic Games, you play every second day. If you finish top of your group you have a 3-day break before the quarter finals (World Champs) or semifinals (Olympic Games). On your rest days you are allocated two 1-hr training sessions and the day of your game a single 1-hr training session. The game-day training session is not always used. Sometimes the travel to and from the training venue does not allow you much down time prior to the game, so you weigh the positives and negatives of having that session.

These sessions are normally broken up into the following:

- During the first 40 min the team is broken into two groups. For 20 min one group does sprints and recovery and the other group passes and shoots; then the groups change activities for the next 20 min.

- During the last 20 min, the team does tactical work, working on areas we need to address before playing our next opponent.

Recovery

Throughout the preparatory training phase and lead-up competition phase and during the major competition, recovery plays an important part in our programming. We believe that proper recovery helps the athletes push through the highly intensive training phases while maintaining an intense work ethic.

Throughout this time we use various recovery techniques including ice baths (two or three times a week), hot and cold contrast baths, massage, compression stockings, and individual and partner stretching. During the World Championships, three portable ice baths were installed on the balconies of the players' rooms and were compulsory after every game. These were also available to the players whenever they needed them individually.

Nutrition and Strength Training

From 2005 through 2007 we emphasized producing stronger and bigger athletes, which required both good nutrition and strength training. Nutritionists constantly monitored and advised our athletes on improving their recovery and performance through nutrition, including the timing of food and fluid intake. This allowed the players to gain and maintain weight during heavy training phases.

We believe that strength training is so important that we have both sacrificed other areas and have worked to change the players' attitudes about strength training. We have set strength goals for all individuals, and players who have not reached their targets, as indicated by regular strength testing, have been excluded from teams and squads.

Over the past 3 years the athletes have had a maximum of 3 weeks off strength training. One has been for the last week of games during a major competition and the other two have been during complete rest from all training. Thus, during the 2005 and 2007 World Championships we continued our strength training for the first week of competition. During the lead-up to competition phase and major competition phase, the focus of our training was more on power than on strength.

Psychology and Group Dynamics

Because we do not have a very high budget we have to prioritize what areas to train given the time the whole squad is together. Over the last 3 years we have chosen these areas to focus on: mental toughness, communication in the water, job focus, self-belief, and confidence.

Our team sport psychologist attends 2 to 4 days of a 1-week or 10-day training camp within Australia and attends major international competitions in Australia such as the 2007 World Championships. During the World Championships the team had a 30- to 45-min sport psychology session every 3 or 4 days and athletes had individual sessions when necessary. In general the group dynamics within the team have been very good. We monitor this through individual discussions with the players while also encouraging the team leaders to take more responsibility for the team's welfare and behavior. This area is continually addressed by the team psychologist with the leadership group.

Training Programs and Training Loads

To avoid underperforming when performance really counts, it is vital to have a periodization plan leading up to the taper for the culminating event. Following is the periodization plan we used early in the team's training as well as for the lead-up program for 2007 World Championships.

Periodization Plan

Table 12.4 shows the training loads and periodization plan for the 2007 World Championships. This preparation started 20 weeks out from the World Championships, when the World Championships squad of 20 players was chosen.

Table 12.4 Australian Women's Water Polo Team World Championships Preparation Plan

		30-Oct-06	6-Nov-06	13-Nov-06	20-Nov-06	27-Nov-06	4-Dec-06	11-Dec-06	18-Dec-06	25-Dec-06	1-Jan-07	8-Jan-07	15-Jan-07	22-Jan-07	29-Jan-07	5-Feb-07	12-Feb-07	19-Feb-07	26-Feb-07	5-Mar-07	12-Mar-07	19-Mar-07	26-Mar-07
Competitions	Domestic				✓	✓		✓	✓		✓	✓	✓										
	International						✓										✓		✓		✓	✓	✓
	Location				National League season	National League season	Holiday Cup	National League season	National League season		National League season	National League season	National League season	Camp	Camp	Camp	Thetis Cup	Camp with Greece	Barcelona Tournament	Return home	Sydney Tournament	World Championships	World Championships
Camps														✓	✓	✓		✓					
Testing/medical control																							
Training week		1	2	3	4	5	6	7	8	9	10	11	12	13	14	15	16	17	18	19	20	21	22
Weeks to go		20	19	18	17	16	15	14	13	12	11	10	9	8	7	6	5	4	3	2	1	0	0
Training load		L	M	H	H	L	H	M	H	L	H	H	L	H	H	L	H	M	H*	L**	M***	C	C
Training phase		General preparation												Specific preparation							Taper	Comp	
Skills	Technical																						
	Tactical																						
Physiology	Aerobic capacity			Develop										Maintain									
	Aerobic power																						
	Anaerobic capacity																						
	Anaerobic power																						
	Repeat effort			Prepare										Develop									
	Speed			Develop →																			
	Active recovery																						

Comp = Competition

The training load row: L = Low, M = Medium, and H = Heavy

*Key week to overload as it sets the team up for the taper. Intensity is more important than volume, but with games and hard training sessions, some volume is inevitable.

**Recovery from flight and jet lag is vital.

***Low overall volume, but quality and intensity must be emphasized.

After selecting 20 players for the national World Championships squad in October, we met with the National League clubs (to which all the players belonged) and decided to break the season into three sections.

- The first section would last for 5 weeks and run from November 20 to December 23.
- The second section would go until 3 weeks after the New Year, running from January 3 to January 21.
- The third section would be after the World Championships, starting April 10 and going to May 12.

This would allow us an 8-week lead-up preparation phase till the start of the World Championships.

Table 12.5 World Championships Lead-Up*

November		December		January		February		March	
1		**1**	AUS v USA	1	Training	1	AIS	1	Easy day
2		2	AUS v NED	2		**2**		**2**	AUS v HUNG
3		**3**		3	with	3	Training	3	AUS v SPAIN
4		4	USA	4		4		**4**	AUS v GRE
5		**5**	AUS v ITALY	5	SIS/SAS	**5**	Camp	5	Travel
6		6	AUS v RUS & CAN	6		6		6	Home
7		**7**	AUS v USA	7	and	**7**		7	Day off
8		8	AUS v NED	8		8		8	Travel
9		**9**	AUS v CAN	9		9	AIS	9	AUS v GRE
10		10	Travel	10		10	Home	10	AUS v SPAIN
11		11	Training	11	National	11	Home	11	AUS v USA & AUS
12		12		12		**12**	Home	12	AUS v ITALY
13		13	with	13	League	13	Travel	**13**	Day off
14		14		14		14	Easy day	14	Day off
15		15	SIS/SAS	15	clubs	**15**	TRA v GRE	15	TRA v USA
16		16		16		16	AUS v RUSSIA	**16**	WGTS & TRAVEL
17		17	and	17		**17**	AUS v CAN	17	Day off
18		18		18	set	18	AUS v GRE	**18**	WORLD CHAMPS
19		19	National	19		19	TRA v GRE	19	AUS v CAN
20	Training	20		20		**20**	TRA v GRE	20	Rest day
21	with	21	League clubs	21	progams	21	EASY DAY	21	AUS v PUERTO RICO
22		22		**22**		**22**	TRA v GRE	22	Rest day
23	SIS/SAS	23	set programs	23	GK, CB, & CF	23	TRA v GRE	23	AUS v BRAZIL
24	and	24		**24**	Camp	**24**	TRA v GRE	24	Rest day
25		25		25	Perth	25	Travel	25	TRA v USA
26	National	26		26	Day off	**26**	TRA v SPAIN	26	Rest day
27	League clubs	27		27	Day off	27	TRA v SPAIN	27	AUS v ITALY
28		28		28	Day off	**28**	TRA v SPAIN	28	Rest day
29	set programs	29		**29**	STR testing	29		29	AUS v RUSSIA
30	Travel	30		30	Swim test	30		30	Rest day
31		31		**31**	Camp	31		31	AUS v USA

*To properly interpret the table, please note the following definitions:

Day off = only a weight session in the morning; **Weight training days (days in bold)** = weight training plus games or other training; **Rest day** = no games at world championships, but 2 training sessions; **Easy day** = light or recovery session; **Travel** = flights to other countries or cities; **Home** = time at home; may be required to do weights.

Also note the following abbreviations: **AIS** = Australian Institute of Sport; **CAN** = Canada; **CB** = center back; **CF** = center forward; **GK** = goal keepers; **GRE** = Greece; **HUNG** = Hungary; **NED** = Netherlands; **RUS** = Russia; **SAS** = State Academy of Sport; **SIS** = State Institute of Sport; **TRA** = Training Match

Lead-Up to the World Championships

The day-to-day lead-up to the World Championships is described in table 12.5, including training camps and games, tournaments, and rest days. Because of the size of Australia and with the National League season being played for 7 1/2 weeks during our general and specific preparation time, we had to train in three separate venues. Our 20 players were based in these centers, in Sydney, Brisbane, and Perth. To help prepare we also had to incorporate the National League club training with the national squad programs. Each club that a national squad player was linked with had a different program based on when their National League team was scheduled to play their games both at home and away.

The National League season's break after January 22 allowed us to prepare together with training camps in Australia or overseas and lead-up tournaments. The downside of this was that during 10 weeks the athletes were only at home for a total of 7 days.

World Championships Review

Although all the athletes were in good shape prior to Christmas, some of that fitness was lost during the Christmas–New Year break, and the second section of the National

League season had not proved to be as beneficial as we had hoped. This meant that we had to work harder than originally planned and extra fitness sessions were needed during the 6 weeks between the National League finish and the beginning of the taper. The other area of concern was that four of our athletes sustained injuries. Fortunately, two of these injured players overcame their injuries to be selected to the final team.

Our first game (against Canada) was one of the heaviest challenges of the World Championships. Although only eight exclusions were called during the game, we were able to come from behind four times to win 5–4. A very good performance by the Canadian goalie and our own nerves caused by playing in front of a big home crowd for the first time seemed to affect our ability to score.

We won the next two games quite comfortably, giving us a straight run through to the quarter finals, which gave us a *bye* and a 3-day break before our quarter final against the winner of Italy versus The Netherlands. This meant that we had not had a hard game for 9 days, since our opening game against Canada. We didn't want to go into the quarterfinals underdone, and fortunately neither did the United States, so we organized a training match on the day of the elimination games for the quarter finals. The training game was played at the same speed as, and at a higher level than, our other two round games during the World Championships.

In the quarter final the training game against the United States proved very beneficial. Our team played extremely well and dominated the opposition. The semifinal with the Russians featured several exciting reverses, but in the end we won 12–9. This took us to the gold medal match and the chance to defend our number one ranking earned at the 2006 World Cup. Our opponent, however, was the United States, whom we had not beaten since August 2005. We started strongly and battled back and forth throughout the game, but in the end, the United States won 6–5 by scoring the winning goal in the last minute of the game. Even though we failed to win the tournament, we proved that we were one of the best teams in the world and were in a good position in the lead-up to Beijing.

After all the heavy training, National League commitments, and training camps, the 2 weeks of tapering allowed us to recover and fine-tune the players' fitness, technical skills, and tactical preparation to achieve their goals during the tournament. Prior to the tapering phase our performances were a bit inconsistent in the Thetis cup in Athens, Madrid Open, and Sydney Cup. But as we went through our tapering phase, our performances improved gradually and resulted in our playing our best water polo in the major games, which allowed us to make the gold medal match. We had become a lot sharper and I believe one of the fastest and fittest teams in the competition.

Applying Greg McFadden's Expert Advice   AT A GLANCE

Greg McFadden explains to the reader how he and his team prepared for the 2007 World Championships, a tournament that finished with a silver medal in a very close final game. Key ideas include maintaining a high intensity, increasing the specificity, decreasing the duration of the sessions, and adequately quantifying the individual training load in the lead-up tournaments preceding the major event; making sure that players who are in the water longer during lead-up tournaments receive enough recovery, whereas those receiving less match time receive extra training to maintain their fitness; using posttraining recovery techniques and optimal nutrition strategies; emphasizing players' body size, strength, and power, because this is considered to have a major impact on game quality; and periodizing the training plan to achieve peak performance at the desired time.

It is interesting to note that Ric Charlesworth, the Springboks coaching staff, and Greg McFadden all mention the importance of taking a few days off training during the final 2 weeks prior to the commencement of the major tournament, with the aim of recovering physically and psychologically from the heavy training loads and preparing for the upcoming major tournament. This seems to be a common strategy for winning teams!

Dragan Matutinovic

Securing Silver in Olympic Men's Water Polo

The first and most important thing that came to mind when I started to think about preparing for the Olympic Games in Barcelona as the head coach of the Spanish men's water polo team was the final objective, what we wanted to achieve considering the quality of the squad, on an individual basis and also as a team. The final objective was no other than to play the final of the Olympic Games. Perhaps it was a quite daring objective, because until then Spain had never won an Olympic medal, but I believed in my players, I believed in their capabilities, and I believed they could do it.

During the Olympic year of 1992, everything was focused on that great objective. We had 5 months to prepare, and we put together a long and detailed plan for that time.

Preparation

For the first phase of the plan, we headed to Andorra for 3 weeks with about 30 players, to start a very hard preparation. The program consisted of 8 to 10 hr of training per day for 2 days, followed by a day of active rest (e.g., a football game). It was not easy to cope either physically or psychologically with the demands of the program. But only with such an amount of work would every team member achieve flawless physical fitness, enabling the players to respond adequately during the most psychologically difficult moments of a match.

DRAGAN MATUTINOVIC

Dragan Matutinovic has coached elite water polo at both club and national team levels for more than 20 years. At club level, he has won two European Cups and a number of national leagues in various European countries. At national team level, he has won silver and bronze medals at European and World Championships and also the Olympic Games. Dragan is now the head of player development of the Croatian Water Polo Federation.

Coach Dragan Matutinovic shouts to his team during the European water polo final four in 2001. In the 1992 Olympic games, Matutinovic led the Spanish men's water polo team to win the silver medal using strategies that began with intense physical training and ended with enabling the athletes to remain focused yet relaxed.

© AP Photo/Bozidar Vukicevic

From the very first day I started to set targets for the players, for example, 1 hr of jogging in the morning: The target was to reach the top of a nearby mountain by the last day of the winter preparation. The players knew how far they had to go by the last day. They were eager to achieve this goal, so each day they would cover more and more kilometers, until they all reached the target that had been set, a target just as difficult as playing the Olympic final. The training volume for swimming was also huge, reaching up to 10,000 m a day swum at solid pace. They managed to do this too, setting phenomenal swimming times and attaining an amazing level of physical endurance, which was the foundation for the objective they were pursuing.

After finishing all that hard strength and swimming fitness training, we moved on to improving the tactical components of the game, to creating the type of game we would play during the Olympic Games. We played lots of friendly matches and tournaments all over the world against national teams that played very different types of water polo. During that time, we never stopped training our strength and endurance. All tactical variants were repeated until we reached perfection. The team's discipline was impeccable, and I made it my goal to motivate the players every day so that they would give the maximum of themselves.

At the end of the 5 months and after having gone through all the preparation phases, we picked our 13 players from the initial group of 30: those who were prepared physically and psychologically to face all the difficulties implied by a tournament like the Olympic Games, in which we had to play seven matches in 7 days.

The Games

And so the 13 best-prepared players entered the Olympic Village. After all those running and swimming kilometers, all those trips and matches, all those hours and hours perfecting all the tactical variations (extra man, man down, positional offense, counterattack, defense), we came to the most important 7 days, the week prior to the Games when we had to forget all the previous suffering. We had to concentrate and motivate for the beginning of our Olympic Games.

An additional difficulty came from the fact that because we were playing in Barcelona, the pressure was huge and so were the expectations. During those final 7 days, we coaches had to prepare the players for the Olympic tournament and for the psychological effect of possible defeats during the first phase. Our role was to relax the players during the days before the tournament so that they could play free of pressure, relaxed but concentrated, instead of thinking they had to do something. That is when the work of Dr. Miguel Masgrau was extremely helpful. Using acupuncture and other alternative methods he helped the players be prepared physically and mentally, so the Games could begin. We were prepared to perform, to face all the requirements of such a competition and attain the result we had been aiming for, the final of the Olympic Games.

Applying Dragan Matutinovic's Expert Advice | AT A GLANCE

In this section Dragan Matutinovic describes a successful approach to a team's preparation for a major tournament, in this case the 1992 Olympic Games. In his section, Matutinovic proposes a training methodology that reflects some of the training methods used in the former Eastern block countries. Nevertheless, there are some common concepts in this and the previous reports, such as the importance of establishing a clear goal from the outset and believing it could be achieved. Key ideas include setting specific, difficult targets for the players to increase their physical and mental strength and determination; creating a playing style early in the preparation process and competing against teams that played very different types of games; and helping the players to stay motivated, focused, relaxed, and free of external and internal pressure in the days before and during the event.

And that is exactly what we did! Not only did those players win an Olympic silver medal (after losing the final to Italy in overtime), but every single one of them continued to play for the national team and won several more medals at international tournaments in the following years, including World and European Championships and Olympic Games. It is my belief that the foundation of the preparation for the 1992 Barcelona Olympic Games remained with the players during the following competitions, because it had been so hard that anything that came after that became much easier to cope with.

The most important thing that I must underline is that all this would have been impossible without the incredible talent and quality of the players. I am still proud of all my players, all my assistants, the federation which helped us so much, and all those who helped us during that time, because without them none of this would have been possible.

ADP rephosphorylation   Addition of a high-energy phosphate group to a molecule of adenosine diphosphate (ADP) to form ATP.

aldosterone   Mineralocorticoid steroid hormone secreted by the adrenal cortex that prevents dehydration by stimulating sodium and water absorption at the kidneys.

anabolic hormone   Testosterone-like hormone that stimulates growth by increasing protein synthesis.

androgen   Natural or synthetic compound that stimulates or controls the growth, development, and maintenance of male sex characteristics.

androgenic–anabolic activity   Effects of the hormone testosterone: promoting and controlling growth, spermatogenesis, and maintenance of male sex characteristics.

ATP hydrolysis   Breakdown of adenosine triphosphate (ATP) to adenosine diphosphate (ADP) and inorganic phosphate (P_i), with addition of elements of water and releasing a large amount of energy.

autocrine   Mode of chemical messenger action in which the messenger binds to receptors on the cell that secreted it, affecting the function of the secretory cell itself.

B-cell lymphocyte   Lymphocyte that upon activation proliferates and differentiates into plasma cells that produce and release antibodies.

biphasic response   Two differentiated and distinct responses that are separated in time and characterized by an immediate reaction, a time of quiescence, and a recurrent reaction.

blood lactate concentration   Concentration in the blood of a three-carbon molecule formed when lactic acid produced by anaerobic glycolysis dissociates to lactate and hydrogen ions.

blood lactate–swimming velocity curve   Curve describing the evolution of blood lactate concentration in response to increasing swimming velocities.

Borg's Rating of Perceived Exertion (RPE)   A numerical scale for rating perceived exertion.

bye   In a tournament, the position of a player or team not paired with a competitor in an early round and thus automatically advanced to play in the next round.

cardiac index   Cardiodynamic parameter that relates cardiac output to body surface area, thus relating performance of the heart to the size of the individual.

cardiac output   Volume of blood ejected by the left ventricle each minute.

catabolic hormone   Hormone that stimulates the cellular breakdown of complex organic molecules.

catecholamine   Biologically active amines (organic compounds derived from ammonia), such as dopamine, epinephrine, and norepinephrine, all of which have similar chemical structures and powerful effects similar to those of the sympathetic nervous system.

cortisol   Main corticosteroid hormone released by the adrenal cortex; regulates various aspects of organic metabolism, including stimulation of gluconeogenesis, increased mobilization of free fatty acids, decreased use of glucose, and stimulated catabolism of proteins.

C-peptide   Peptide that is made when proinsulin is released from the pancreas into the blood and then split into insulin and C-peptide, the level of which is a gauge of how much insulin is being produced in the body.

creatine kinase   Muscle enzyme that facilitates the breakdown of phosphocreatine to creatine and inorganic phosphate; is occasionally increased in the blood following strenuous or eccentric exercise, most probably as a result of altered permeability of tissue cell membranes.

cytokines   Protein intercellular messengers secreted by macrophages, monocytes, lymphocytes, and other cells that influence cells of the immune system.

deformability   The ability of cells, such as red blood cells, to change shape as they pass through narrow spaces such as the microvasculature.

diastolic cavity   Interior chamber of the heart during the relaxation phase of the cardiac cycle.

economy of movement   Energy cost of moving the body at a given submaximal exercise intensity.

erythrocyte superoxide dismutase   In red blood cells, enzyme that catalyzes the destruction of superoxide anions to hydrogen peroxide, playing a critical role in the defense of cells against the toxic effects of oxygen radicals.

erythropoiesis   Erythrocyte (i.e., red blood cell) production.

eumenorrheic   Having normal menstruation.

extravascular hemolysis   Breakdown of red blood cells with release of hemoglobin in the spleen and liver.

ferritin   Iron-binding protein that stores iron in the body.

fractional shortening   Measure of left ventricular performance; it measures and calculates a ratio of the change in the diameter of the left ventricle between the contracted and relaxed states.

free radical scavenging   Action of antioxidant chemical substances that protect body cells from the damaging effects of free radicals.

glutathione   Antioxidant amino acid produced in the liver to protect body cells from the damaging effects of free radicals.

glycoprotein   Protein containing a relatively small carbohydrate group attached to a large protein.

haptoglobin   A glycoprotein that binds free hemoglobin released into the circulation to conserve body iron.

hematocrit   Percentage of total blood volume occupied by blood cells or formed elements.

hematopoiesis   Blood cell formation and differentiation.

hemodilution   An increase in blood plasma that results in a dilution of the blood's cellular contents.

hemoglobin   Protein composed of four globular polypeptide chains, each bound to a single molecule of heme, located in red blood cells and responsible for the transport of most blood oxygen.

hemolysis   Breakdown (lysis) of red blood cells with liberation of hemoglobin.

hemolytic condition   State is which there is a predominance of red blood cell breakdown over red blood cell production.

hypervolemia   An abnormal increase in the volume of blood plasma.

immunoglobulin   A circulating antibody, of which there are five classes: IgA, IgD, IgE, IgG, IgM.

insulin-like growth factor-I   Peptide that has growth-promoting effect by mediating the mitosis-stimulating effect of growth hormone.

intramuscular creatine phosphate supercompensation Hypothetical elevation of creatine phosphate concentration inside the muscle cells beyond baseline values.

intravascular hemolysis Breakdown of red blood cells within the blood vessels.

J-curve Model describing the relationship between exercise workload and infection, which suggests that moderate exercise training may decrease the frequency of infections whereas excessive, exhausting exercise can lead to the opposite, a situation that has been described by a J-curve.

lactate recovery curves Evolution of blood lactate concentration during recovery from exercise.

leptin A protein hormone that plays a key role in regulating energy intake and energy expenditure, including appetite and metabolism.

leukocyte White blood cell.

lipoperoxidation Oxidative degradation of lipids occurring when free radicals capture electrons from the lipids in cell membranes, resulting in cell damage.

luteal phase Latter half of the menstrual cycle, which begins the day after ovulation with the formation of the corpus luteum and ends in either pregnancy or luteolysis.

luteinizing hormone Anterior pituitary hormone that in females assists follicle-stimulating hormone in follicle stimulation, triggers ovulation, and promotes the maintenance and secretion of endometrial glands; in males, it stimulates spermatogenesis.

lymphocyte A leukocyte of the lymphatic system that is responsible for specific immune defenses.

mean corpuscular hemoglobin The average mass of hemoglobin per red blood cell in a sample of blood, calculated by dividing the total mass of hemoglobin by the number of red blood cells in a volume of blood.

mechanical fragmentation Breakdown of red blood cells by the continuous buffeting in the circulation.

meta-analysis A statistical procedure in which the results of several studies are pooled together and analyzed as if they were the results of one large study.

mitochondrial capacity Capacity of mitochondria to generate adenosine triphosphate through oxidative phosphorylation.

mitochondrial enzymes Biomolecules, generally proteins, that catalyze chemical reactions inside mitochondria, which aerobically generate most of the cell's supply of adenosine triphosphate.

muscle glycogen A highly branched polysaccharide composed of a long chain of glucose molecules that is the major form of carbohydrate storage in the body.

muscle oxygenation Amount of oxygen that is present in the muscle.

neuroendrocrine fatigue Fatigue associated with dysfunction of the central nervous system and the endocrine system.

neutrophil Phagocytic microphage that is very numerous and usually the first of the mobile phagocytic cells to arrive at an area of injury or infection, also releasing chemicals involved in inflammation.

O_2 pulse Amount of oxygen uptake per heartbeat, used as a measure of cardiovascular efficiency.

open window Time of impaired immunological function following strenuous exercise that increases the risk of subclinical and clinical infection.

osmotic resistance Degree of resistance of red blood cells to a decrease of the salt content of their environment.

perception of effort   Stress a subject perceives when performing physical exercise.

periodization   Organized approach to training that involves progressive cycling of various aspects of a training program during a specific amount of time.

plasma   Liquid portion of blood.

Profile of Mood States (POMS)   A 65-item Likert-format questionnaire that provides a measure of total mood disturbance and six specific mood states (tension, depression, anger, vigor, fatigue, confusion).

progressive taper   A taper in which the training load is reduced in a gradual manner, as opposed to a sudden constant reduction.

prolactin   Peptide hormone secreted by the anterior pituitary gland that stimulates the functional development of the mammary gland in females; prolactin release is sometimes used as an indirect measure of the neurotransmitter 5-hydroxytryptamine activity.

Recovery–Stress Questionnaire for Athletes   An instrument that systematically reveals the recovery–stress state of athletes, indicating the extent to which they are physically or mentally stressed, whether they are capable of using individual strategies for recovery, and which strategies are used.

reduced training   A nonprogressive, standardized reduction in the quantity of training. This procedure has also been referred to as "step taper."

renin   Hormone secreted by the kidneys that acts as an enzyme that catalyzes conversion of a plasma protein called angiotensinogen into angiotensin I, thus participating in the regulation of extracellular volume and arterial vasoconstriction.

respiratory exchange ratio (RER)   Ratio between net output of carbon dioxide and the simultaneous net uptake of oxygen.

reticulocyte   Immature red blood cell.

serum   Blood plasma from which fibrinogen and other clotting proteins have been removed as a result of clotting.

stroke volume   Volume of blood ejected by the left ventricle of the heart in one contraction.

submaximal energy expenditure   Amount of calories expended to perform an exercise of submaximal intensity.

substrate utilization   Use of carbohydrate, fat, and protein as a fuel source.

Super 14 club competition   The largest rugby union club championship in the southern hemisphere, consisting of four teams from Australia, five from New Zealand, and five more from South Africa.

systolic cavity   Interior chamber of the heart during the contraction phase of the cardiac cycle.

T-cell   Lymphocyte derived from precursor that differentiated in the thymus; responsible for cellular immunity and the coordination and regulation of the immune response.

taper   A reduction of the training load during several days prior to a major competition.

testosterone   The predominant testicular androgen or male sex hormone, essential for spermatogenesis as well as growth, maintenance, and development of reproductive organs and secondary sexual characteristics of males.

thyroxine   Iodine-containing amine hormone secreted by the thyroid gland that increases the rate of cellular metabolism and contractility of the heart.

transferrin   Iron-binding protein carrier for iron in plasma.

triiodothyronine Iodine-containing amine hormone secreted by the thyroid gland that increases the rate of cellular metabolism and the rate and contractility of the heart.

Type I muscle fibers Muscle fibers that have a high oxidative and a low glycolytic capacity, associated with endurance-type activities.

Type II muscle fibers Muscle fibers that have a low oxidative capacity and a high glycolytic capacity, associated with speed or power activities.

vasopressin Peptide hormone synthesized in the hypothalamus and released from the posterior pituitary that increases water permeability of the kidney's collecting ducts; also called antidiuretic hormone (ADH).

ventilatory function Measure of dynamic lung volumes and maximal air flow rates.

ventilatory threshold Occurrence during exercise of progressively increasing intensity at which there is a nonlinear increase in ventilation.

ventricular septal wall Wall dividing the left and right ventricles of the heart.

$\dot{V}O_2max$ Maximal capacity for oxygen consumption by the body during maximal exertion.

Achten J, Jeukendrup AE. Heart rate monitoring: applications and limitations. Sports Med 2003; 33: 517-538.

Adlercreutz H, Härkönen M, Kuoppasalmi K, et al. Effect of training on plasma anabolic and catabolic steroid hormones and their response during physical exercise. Int J Sports Med 1986; 7: 27-28.

Armstrong LE. Nutritional strategies for football: counteracting heat, cold, high altitude and jet-lag. J Sport Sci 2006; 24: 723-740.

Atlaoui D, Duclos M, Gouarne C, et al. The 24-hr urinary cortisol/cortisone ratio for monitoring training in elite swimmers. Med Sci Sports Exerc 2004; 36: 218-224.

Avalos M, Hellard P, Chatard JC. Modeling the training-performance relationship using a mixed model in elite swimmers. Med Sci Sports 2003; 35: 838-846.

Baj Z, Kantorski J, Majewski E, et al. Immunological status of competitive cyclists before and after the training season. Int J Sports Med 1994; 15: 319-324.

Balague G. Periodization of psychological skills training. J Sci Med Sports 2000; 3: 230-237.

Bangsbo J. Préparation physique en vue de la Coupe du monde de football. Science & Sport 1999; 14: 220-226.

Bangsbo J, Mohr M, Poulsen A, Perez-Gomez J, Krustrup P. Training and testing the elite athlete. J Exerc Sci Fit 2006; 4: 1-14.

Banister EW, Calvert TW. Planning for future performance: implications for long term training. Can J Appl Sport Sci 1980; 5: 170-176.

Banister EW, Calvert TW, Savage MV, et al. A systems model of training for athletic performance. Aust J Sports Med 1975; 7: 57-61.

Banister EW, Carter JB, Zarkadas PC. Training theory and taper: validation in triathlon athletes. Eur J Appl Physiol 1999; 79: 182-191.

Banister EW, Fitz-Clarke JR. Plasticity of response to equal quantities of endurance training separated by non-training in humans. J Therm Biol 1993; 18: 587-597.

Banister EW, Hamilton CL. Variations in iron status with fatigue modelled from training in female distance runners. Eur J Appl Physiol 1985; 54: 16-23.

Banister EW, Morton RH, Fitz-Clarke J. Dose/response effects of exercise modeled from train- ing: physical and biochemical measures. Ann Physiol Anthrop 1992; 11: 345-356.

Berger BG, Grove JR, Prapavessis H, et al. Relationship of swimming distance, expectancy, and performance to mood states of competitive athletes. Percept Mot Skills 1997; 84: 1199-1210.

Berger BG, Motl RW, Butki BD, et al. Mood and cycling performance in response to three weeks of high-intensity, short-duration overtraining, and a two-week taper. Sport Psychol 1999; 13: 444-457.

Berglund B, Säfström H. Psychological monitoring and modulation of training load of world-class canoeists. Med Sci Sports Exerc 1994; 26: 1036-1040.

Bessman JD, Ridgeway Gilmer P, Gardner FH. Improved classification of anemias by MCV and RDW. Am J Clin Pathol 1983; 80: 322-326.

Best R, Walker BR. Additional value of measurement of urinary cortisone and unconjugated cortisol metabolites in assessing the activity of 11 beta-hydroxysteroid dehydrogenase in vivo. Clin Endocrinol (Oxf) 1997; 47: 231-236.

Bishop D, Edge J. The effects of a 10-day taper on repeated-sprint performance in females. J Sci Med Sport 2005; 8: 200-209.

Bonifazi M, Sardella F, Luppo C. Preparatory versus main competitions: differences in performances, lactate responses and pre-competition plasma cortisol concentrations in elite male swimmers. Eur J Appl Physiol 2000; 82: 368-373.

Borg G. Perceived exertion as an indicator of somatic stress. Scand J Rehab Med 1970; 2: 92-98.

Borg G. Psychophysical bases of perceived exertion. Med Sci Sports Exerc 1982; 14: 377-381.

Borg G, Hassmen P, Lagerstrom M. Perceived exertion related to heart rate and blood lactate during arm and leg exercise. Eur J Appl Physiol 1987; 65: 679-685.

Bosquet L, Montpetit J, Arvisais D, et al. Effects of tapering on performance: a meta-analysis. Med Sci Sports Exerc 2007; 39: 1358-1365.

Bothwell TH, Charlton RW, Cook JD, et al. Iron Metabolism in Man. Oxford, UK: Blackwell Scientific, 1979.

Brodthagen UA, Hansen KN, Knudsen JB, et al. Red cell 2,3-DPG, ATP, and mean cell volume in highly trained athletes. Effect of long-term submaximal exercise. Eur J Appl Physiol 1985; 53: 334-338.

Bruunsgaard H, Hartkopp A, Mohr T, et al. In vivo cell-mediated immunity and vaccination response following prolonged, intense exercise. Med Sci Sports Exer 1997; 29: 1176-1181.

Bunt JC. Hormonal alterations due to exercise. Sports Med 1986; 3: 331-345.

Burke ER, Falsetti HL, Feld RD, et al. Blood testing to determine overtraining in swimmers. Swimming Tech 1982a; 18: 29-33.

Burke ER, Falsetti HL, Feld RD, et al. Creatine kinase levels in competitive swimming during a season of training. Scand J Sports Sci 1982b; 4: 1-4.

Busso T. Variable dose-response relationship between exercise training and performance. Med Sci Sports Exerc 2003; 35: 1188-1195.

Busso T, Benoit H, Bonnefoy R, et al. Effects of training frequency on the dynamics of performance response to a single training bout. J Appl Physiol 2002; 92: 572-580.

Busso T, Candau R, Lacour JR. Fatigue and fitness modelled from the effects of training on performance. Eur J Appl Physiol 1994; 69: 50-54.

Busso T, Denis C, Bonnefoy R, et al. Modeling of adaptations to physical training by using a recursive least squares algorithm. J Appl Physiol 1997; 82: 1685-1693.

Busso T, Häkkinen K, Pakarinen A, et al. A systems model of training responses and its relationship to hormonal responses in elite weight-lifters. Eur J Appl Physiol 1990; 61: 48-54.

Busso T, Häkkinen K, Pakarinen A, et al. Hormonal adaptations and modelled responses in elite weighlifters during 6 weeks of training. Eur J Appl Physiol 1992; 64: 381-386.

Busso T, Thomas L. Using mathematical modelling in training planning. Int J Sports Physiol Perf 2006; 1: 400-405.

Butterfield GE, Gates J, Fleming S, Brooks GA, Sutton JR, Reeves JT. Increased energy intake minimises weight loss in men at high altitude. J Appl Physiol 1992; 72: 1741-1748.

Calvert TW, Banister EW, Savage MV, et al. A systems model of the effects of training on physical performance. IEEE Trans Syst Man Cybern 1976; 6: 94-102.

Carli G, Martelli G, Viti A, et al. Modulation of hormone levels in male swimmers during training. In: Biomechanics and Medicine in Swimming. Champaign, IL: Human Kinetics, 1983: 33-40.

Casoni I, Borsetto C, Cavicchi A, et al. Reduced hemoglobin concentration and red cell hemoglobinization in Italian marathon and ultramarathon runners. Int J Sports Med 1985; 6: 176-179.

Cavanaugh DJ, Musch KI. Arm and leg power of elite swimmers increase after taper as measured by biokinetic variable resistance machines. J Swimming Research 1989; 5: 7-10.

Chatard JC, Paulin M, Lacour JR. Postcompetition blood lactate measurements and swimming performance: illustrated by data from a 400-m Olympic record holder. In: Swimming Science V. Champaign, IL: Human Kinetics, 1988: 311-316.

Child RB, Wilkinson DM, Fallowfield JL. Effects of a training taper on tissue damage indices, serum antioxidant capacity and half-marathon running performance. Int J Sports Med 2000; 21: 325-331.

Clement DB, Sawchuk LL. Iron status and sports performance. Sports Med 1984; 1: 65-74.

Cohen J. Statistical Power Analysis for the Behavioral Sciences. Hillsdale, NJ: Erlbaum, 1988.

Convertino VA, Keil C, Greenleaf JE. Plasma volume, osmolality, vasopressin, and renin activity during graded exercise in man. J Appl Physiol 1981; 50: 123-128.

Convertino VA, Keil C, Greenleaf JE. Plasma volume, renin and vasopressin responses to graded exercise after training. J Appl Physiol 1983; 54: 508-514.

Costill DL, King DS, Thomas R, et al. Effects of reduced training on muscular power in swimmers. Physician Sportsmed 1985; 13: 94-101.

Costill DL, Thomas R, Robergs A, et al. Adaptations to swimming training: influence of training volume. Med Sci Sports Exerc 1991; 23: 371-377.

Coutts A, Reaburn P, Piva TJ. Changes in selected biochemical, muscular strength, power, and endurance measures during deliberate overreaching and tapering in rugby league players. Int J Sports Med 2007a; 28: 116-124.

Coutts AJ, Wallace LK, Slattery KM. Monitoring changes in performance, physiology, biochemistry, and psychology during overreaching and recovery in triathletes. Int J Sports Med 2007b; 28: 125-134.

Cumming DC, Wall SR. Non-sex hormone-binding globulin-bound testosterone as a marker for hyperandrogenism. J Clin Endocrinol Metab 1985; 61: 873-876.

D'Acquisto LJ, Bone M, Takahashi S, et al. Changes in aerobic power and swimming economy as a result of reduced training volume. In: Swimming Science VI. London: E & FN Spon, 1992: 201-205.

Dawkins R. The Selfish Gene (30th Anniversary Edition). Oxford, UK: Oxford University Press, 2006.

Dressendorfer RH, Petersen SR, Moss Lovshin SE, et al. Performance enhancement with maintenance of resting immune status after intensified cycle training. Clin J Sport Med 2002a; 12: 301-307.

Dressendorfer RH, Petersen SR, Moss Lovshin SE, et al. Mineral metabolism in male cyclists during high-intensity endurance training. Int J Sport Nutr Exerc Metab 2002b; 12: 63-72.

Dufaux B, Hoederath A, Streitberger I, et al. Serum ferritin, transferrin, haptoglobin, and iron in middle- and long-distance runners, elite rowers, and professional racing cyclists. Int J Sports Med 1981; 2: 43-46.

Ekstrand J, Walden M, Hagglund M. A congested football calendar and the well being of players: correlation between match exposure of European footballers before the World Cup 2002 and their injuries and performances during that World Cup. Br J Sports Med 2004; 38: 439-497.

Eliakim A, Nemet D, Bar-Sela S, et al. Changes in circulating IGF-I and their correlation with self-assessment and fitness among elite athletes. Int J Sports Med 2002; 23: 600-603.

Ferret JM, Cotte T. Analyse des différences de préparation médico-sportive de l'Equipe de France de football pour les coupes du monde 1998 et 2002. In: Lutter contre le Dopage en gérant la Récupération physique. JC Chatard, Ed. Publications de l'Université de Saint-Étienne, 2003: 23-26. Saint-Étienne, France.

Fitz-Clarke JR, Morton RH, Banister EW. Optimizing athletic performance by influence curves. J Appl Physiol 1991; 71: 1151-1158.

Flynn MG, Pizza FX, Boone JB Jr, et al. Indices of training stress during competitive running and swimming seasons. Int J Sports Med 1994; 15: 21-26.

Fry RW, Morton AR, Crawford GPM, et al. Cell numbers and in vitro responses of leucocytes and lymphocyte subpopulations following maximal exercise and interval training sessions of different intensities. Eur J Appl Physiol 1992; 64: 218-227.

Fry RW, Morton AR, Keast D. Overtraining in athletes: an update. Sports Med 1991; 12: 32-65.

Galbo H. The hormonal response to exercise. Diabet/Metab Rev 1986; 1: 385-408.

Gibala MJ, MacDougall JD, Sale DG. The effects of tapering on strength performance in trained athletes. Int J Sports Med 1994; 15: 492-497.

Gledhill N. Blood doping and related issues: a brief review. Med Sci Sports Exerc 1982; 14: 183-189.

Gledhill N. The influence of altered blood volume and oxygen transport capacity on aerobic performance. Exerc Sports Sci Rev 1985; 13: 75-94.

Gleeson M. Mucosal immunity and respiratory illness in elite athletes. Int J Sports Med 2000; 21: S33-S43.

Gleeson M, McDonald WA, Cripps AW, et al. The effect of immunity of long-term intensive training in elite swimmers. Clin Exp Immunol 1995; 102: 210-215.

Gleeson M, McDonald WA, Pyne DB, et al. Salivary IgA levels and infection risk in elite swimmers. Med Sci Sports Exerc 1999; 31: 67-73.

Gleeson M, McDonald WA, Pyne DB, et al. Immune status and respiratory illness for elite swimmers during a 12-week training cycle. Int J Sports Med 2000; 21: 302-307.

Gleeson M, Pyne D. Exercise effects on mucosal immunity. Immunol Cell Biol 2000; 78: 536-544.

Gordon T, Pattullo MC. Plasticity of muscle fiber and motor unit types. Exerc Sport Sci Rev 1993; 21: 331-362.

Graves JE, Pollock ML, Leggett SH, et al. Effect of reduced training frequency on muscular strength. Int J Sports Med 1988; 9: 316-319.

Hague JFE, Gilbert SS, Burgess HJ, et al. A sedentary day: effects on subsequent sleep and body temperatures in trained athletes. Physiol Behav 2003; 78: 261-267.

Hallberg L, Magnusson B. The etiology of "sports anemia." Acta Med Scand 1984; 216: 145-148.

Halson SL, Jeukendrup AE. Does overtraining exist? An analysis of overreaching and overtraining research. Sports Med 2004; 34: 967-981.

Harber MP, Gallagher PM, Creer AR, et al. Single muscle fiber contractile properties during a competitive season in male runners. Am J Physiol Regul Integr Comp Physiol 2004; 287: R1124-R1131.

Hawley J, Burke L. Peak performance: training and nutritional strategies for sport. St. Leonards, NSW, Australia: Allen & Unwin, 1988.

Haykowsky MJ, Smith DJ, Malley L, et al. Effects of short term altitude training and tapering on left ventricular morphology in elite swimmers. Can J Cardiol 1998; 14: 678-681.

Hellard P, Avalos M, Lacoste L, et al. Assessing the limitations of the Banister model in monitoring training. J Sports Sci 2006; 24: 509-520.

Hellard P, Avalos M, Millet G, et al. Modeling the residual effects and threshold saturation of training: a case study of Olympic swimmers. J Strength Cond Res 2005; 19: 67-75.

Hickson RC, Foster C, Pollock ML, et al. Reduced training intensities and loss of aerobic power, endurance, and cardiac growth. J Appl Physiol 1985; 58: 492-499.

Hickson RC, Kanakis C Jr, Davis JR, et al. Reduced training duration effects on aerobic power, endurance, and cardiac growth. J Appl Physiol 1982; 53: 225-229.

Hickson RC, Rosenkoetter MA. Reduced training frequencies and maintenance of increased aerobic power. Med Sci Sports Exerc 1981; 13: 13-16.

Hoffman-Goetz L, Pedersen BK. Exercise and the immune system: a model of the stress response? Immunol Today 1994; 15: 382-387.

Hooper SL, Mackinnon LT, Ginn EM. Effects of three tapering techniques on the performance, forces and psychometric measures of competitive swimmers. Eur J Appl Physiol 1998; 78: 258-263.

Hooper SL, Mackinnon LT, Gordon RD, et al. Hormonal responses of elite swimmers to overtraining. Med Sci Sports Exerc 1993; 25: 741-747.

Hooper SL, Mackinnon LT, Howard A. Physiological and psychometric variables for monitoring recovery during tapering for major competition. Med Sci Sports Exerc 1999; 31: 1205-1210.

Hooper SL, Mackinnon LT, Howard A, et al. Markers for monitoring overtraining and recovery. Med Sci Sports Exerc 1995; 27: 106-112.

Hopkins WG, Hawley JA, Burke LM. Design and analysis of research on sport performance enhancement. Med Sci Sports Exerc 1999; 31: 472-485.

Hopkins WG, Hewson DJ. Variability of competitive performance of distance runners. Med Sci Sports Exerc 2001; 33: 1588-1592.

Hoppeler H. Exercise-induced ultrastructural changes in skeletal muscle. Int J Sports Med 1986; 7: 187-204.

Houmard JA. Impact of reduced training on performance in endurance athletes. Sports Med 1991; 12: 380-393.

Houmard JA, Costill DL, Mitchell JB, et al. Reduced training maintains performance in distance runners. Int J Sports Med 1990a; 11: 46-52.

Houmard JA, Costill DL, Mitchell JB, et al. Testosterone, cortisol, and creatine kinase levels in male distance runners during reduced training. Int J Sports Med 1990b; 11: 41-45.

Houmard JA, Johns RA. Effects of taper on swim performance: practical implications. Sports Med 1994; 17: 224-232.

Houmard JA, Kirwan JP, Flynn MG, et al. Effects of reduced training on submaximal and maximal running responses. Int J Sports Med 1989; 10: 30-33.

Houmard JA, Scott BK, Justice CL, et al. The effects of taper on performance in distance runners. Med Sci Sports Exerc 1994; 26: 624-631.

Ingjer F, Myhre K. Physiological effects of altitude training on young male cross-country. J Sport Sci 1992; 10: 37-47.

Izquierdo M, Ibañez J, González-Badillo JJ, et al. Detraining and tapering effects on hormonal responses and strength performance. J Strength Cond Res 2007; 21: 768-775.

Jeukendrup AE, Hesselink MKC, Snyder AC, et al. Physiological changes in male competitive cyclists after two weeks of intensified training. Int J Sports Med 1992; 13: 534-541.

Johns RA, Houmard JA, Kobe RW, et al. Effects of taper on swim power, stroke distance and performance. Med Sci Sports Exerc 1992; 24: 1141-1146.

Kaiser V, Janssen GME, Van Wersch JWJ. Effect of training on red blood cell parameters and plasma ferritin: a transverse and a longitudinal approach. Int J Sports Med 1989; 10: S169-S175.

Kajiura JS, MacDougall JD, Ernst PB, et al. Immune responses to changes in training intensity and volume in runners. Med Sci Sports Exer 1995; 27: 1111-1117.

Kannus P, Josza L, Renström P, et al. The effects of training, immobilization and remobilization on muskuloskeletal tissue: 1. Training and immobilization. Scand J Med Sci Sports 1992; 2: 100-118.

Kenitzer RF Jr. Optimal taper period in female swimmers. J Swimming Research 1998; 13: 31-36.

Kindermann W. Metabolic and hormonal reactions in overstrain. Semin Orthop 1988; 3: 207-216.

Kirwan JP, Costill DL, Flynn MG, et al. Physiological responses to successive days of intense training in competitive swimmers. Med Sci Sports Exerc 1988; 20: 255-259.

Konig D, Grathwohl D, Weinstock C, et al. Upper respiratory tract infection in athletes: influence of lifestyle, type of sport, training effort and immunostimulant intake. Exer Immunol Rev 2000; 6: 102-120.

Koziris LP, Hickson RC, Chatterton RT, et al. Serum levels of total and free IGF-I and IGFBP-3 are increased and maintained in long-term training. J Appl Physiol 1999; 86: 1436-1442.

Kubukeli ZN, Noakes TD, Dennis SC. Training techniques to improve endurance exercise performances. Sports Med 2002; 32: 489-509.

Kuipers H. Training and overtraining: an introduction. Med Sci Sports Exerc 1998; 30: 1137-1139.

Kuoppasalmi K, Adlercreutz H. Interaction between catabolic and anabolic steroid hormones in muscular exercise. In: Exercise Endocrinology. Berlin: Walter de Gruyter, 1985: 65-98.

Lacour JR, Bouvat E, Barthélémy JC. Post-competition blood lactate concentrations as indicators of anaerobic energy expenditure during 400-m and 800-m races. Eur J Appl Physiol 1990; 61: 172-176.

Lehmann M, Baumgartl P, Wiesenack C, et al. Training-overtraining: influence of a defined increase in training volume vs training intensity on performance, catecholamines and some metabolic parameters in experienced middle- and long-distance runners. Eur J Appl Physiol 1992; 64: 169-177.

Lehmann M, Deickhuth HH, Gendrisch G, et al. Training-overtraining: a prospective experimental study with experienced middle- and long-distance runners. Int J Sports Med 1991; 12: 444-452.

Lehmann M, Foster C, Keul J. Overtraining in endurance athletes: a brief review. Med Sci Sports Exerc 1993; 25: 854-862.

Lowenstein JM. The purine nucleotide cycle revised. Int J Sports Med 1990; 11: S37-S46.

Mackinnon LT. Chronic exercise training effects on immune function. Med Sci Sports Exer 2000; 32: S369-S376.

Mackinnon LT, Hooper S. Mucosal (secretory) immune system responses to exercise of varying intensity and during overtraining. Int J Sports Med 1994; 15: S179-S183.

Mackinnon LT, Hooper SL, Jones S, et al. Hormonal, immunological and hematological responses to intensified training in elite swimmers. Med Sci Sports Exerc 1997; 29: 1637-1645.

Maestu J, Jurimae J, Jurimae T. Hormonal reactions during heavy training stress and following tapering in highly trained male rowers. Horm Metab Res 2003; 35: 109-113.

Mairbäurl H, Humpeler E, Schwaberger G, et al. Training-dependent changes of red cell density and erythrocytic oxygen transport. J Appl Physiol 1983; 55: 1403-1407.

Malisoux L, Francaux M, Theisen D. What do single-fiber studies tell us about exercise training? Med Sci Sports Exerc 2007; 39: 1051-1060.

Malm C. Exercise immunology: a skeletal muscle perspective. Exer Immunol Rev 2002; 8: 116-167.

Manni A, Partridge WM, Cefalu M. Bioavailability of albumin-bound testosterone. J Clin Endocrinol Metab 1985; 61: 705-710.

Margaritis I, Palazetti S, Rousseau A-S, et al. Antioxidant supplementation and tapering exercise improve exercise-induced antioxidant response. J Am Coll Nutr 2003; 22: 147-156.

Martin DT, Andersen MB. Heart rate-perceived exertion relationship during training and taper. J Sports Med Phys Fitness 2000; 40: 201-208.

Martin DT, Andersen MB, Gates W. Using Profile of Mood States (POMS) to monitor high-intensity training in cyclists: group versus case studies. Sport Psychol 2000; 14: 138-156.

Martin DT, Scifres JC, Zimmerman SD, et al. Effects of interval training and a taper on cycling performance and isokinetic leg strength. Int J Sports Med 1994; 15: 485-491.

McCarthy DA, Dale MM. The leucocytosis of exercise: a review and model. Sports Med 1988; 6: 333-363.

McConell GK, Costill DL, Widrick JJ, et al. Reduced training volume and intensity maintain aerobic capacity but not performance in distance runners. Int J Sports Med 1993; 14: 33-37.

McNeely E, Sandler D. Tapering for endurance athletes. Strength Cond J 2007; 29: 18-24.

Millard M, Zauner C, Cade R, et al. Serum CPK levels in male and female world class swimmers during a season of training. J Swimming Research 1985; 1: 12-16.

Millet GP, Groslambert A, Barbier B, et al. Modelling the relationships between training, anxiety, and fatigue in elite athletes. Int J Sports Med 2005; 26: 492-498.

Minors DS, Waterhouse J. Anchor sleep as a synchroniser of rhythms in abnormal routines. Int J Chronobiol 1981; 7, 165-188.

Mondin GW, Morgan WP, Piering PN, et al. Psychological consequences of exercise deprivation in habitual exercisers. Med Sci Sports Exerc 1996; 28: 1199-1203.

Morgan WP, Brown DR, Raglin JS, et al. Psychological monitoring of overtraining and staleness. Br J Sports Med 1987; 21: 107-114.

Morton DP, Gastin PB. Effect of high intensity board training on upper body anaerobic capacity and short-lasting exercise performance. Aust J Sci Med Sport 1997; 29: 17-21.

Morton RH. The quantitative periodization of athletic training: a model study. Sports Med Training Rehabil. 1991; 3: 19-28.

Morton RH, Fitz-Clarke JR, Banister EW. Modeling human performance in running. J Appl Physiol 1990; 69: 1171-1177.

Mujika I. Challenges of team sport research. Int J Sports Physiol Perf 2007a; 2: 221-222.

Mujika I. The influence of training characteristics and tapering on the adaptation in highly trained individuals: a review. Int J Sports Med 1998; 19: 439-446.

Mujika I. Thoughts and considerations for team-sport peaking. Olympic Coach 2007b; 18: 9-11.

Mujika I, Busso T, Lacoste L, et al. Modeled responses to training and taper in competitive swimmers. Med Sci Sports Exerc 1996a; 28: 251-258.

Mujika I, Chatard JC, Busso T, et al. Effects of training on performance in competitive swimming. Can J Appl Physiol 1995; 20: 395-406.

Mujika I, Chatard JC, Busso T, et al. Use of swim-training profiles and performance data to enhance training effectiveness. J Swimming Research 1996b; 11: 23-29.

Mujika I, Chatard J-C, Geyssant A. Effects of training and taper on blood leucocyte populations in competitive swimmers: relationships with cortisol and performance. Int J Sports Med 1996c; 17: 213-217.

Mujika I, Chatard J-C, Padilla S, et al. Hormonal responses to training and its tapering off in competitive swimmers: relationships with performance. Eur J Appl Physiol 1996d; 74: 361-366.

Mujika I, Goya A, Padilla S, et al. Physiological responses to a 6-day taper in middle-distance runners: influence of training intensity and volume. Med Sci Sports Exerc 2000; 32: 511-517.

Mujika I, Goya A, Ruiz E, et al. Physiological and performance responses to a 6-day taper in middle-distance runners: influence of training frequency. Int J Sports Med 2002a; 23: 367-373.

Mujika I, Padilla S. Detraining: loss of training-induced physiological and performance adaptations. Part I. Short-term insufficient training stimulus. Sports Med 2000; 30: 79-87.

Mujika I, Padilla S. Cardiorespiratory and metabolic characteristics of detraining in humans. Med Sci Sports Exerc 2001; 33: 413-421.

Mujika I, Padilla S. Scientific bases for precompetition tapering strategies. Med Sci Sports Exerc 2003a; 35: 1182-1187.

Mujika I, Padilla S. Physiological and performance consequences of training cessation in athletes: detraining. In: Rehabilitation of Sports Injuries: Scientific Basis. WR Frontera, Ed. Malden, MA: Blackwell Science, 2003b: 117-143.

Mujika I, Padilla S, Geyssant A, et al. Hematological responses to training and taper in competitive swimmers: relationships with performance. Arch Physiol Biochem 1997; 105: 379-385.

Mujika I, Padilla S, Pyne D. Swimming performance changes during the final 3 weeks of training leading to the Sydney 2000 Olympic Games. Int J Sports Med 2002b; 23: 582-587.

Mujika I, Padilla S, Pyne D, et al. Physiological changes associated with the pre-event taper in athletes. Sports Med 2004; 34: 891-927.

Neary JP, Bhambhani YN, McKenzie DC. Effects of different stepwise reduction taper protocols on cycling performance. Can J Appl Physiol 2003a; 28: 576-587.

Neary JP, Martin TP, Quinney HA. Effects of taper on endurance cycling capacity and single muscle fiber properties. Med Sci Sports Exerc 2003b; 35: 1875-1881.

Neary JP, Martin TP, Reid DC, et al. The effects of a reduced exercise duration taper programme on performance and muscle enzymes of endurance cyclists. Eur J Appl Physiol 1992; 65: 30-36.

Neary JP, McKenzie DC, Bhambhani YN. Muscle oxygenation trends after tapering in trained cyclists. Dyn Med 2005; 4: 4.

Nehlsen-Cannarella SL, Nieman DC, Fagoaga OR, et al. Saliva immunoglobulins in elite women rowers. Eur J Appl Physiol 2000; 81: 222-228.

Neufer PD. The effect of detraining and reduced training on the physiological adaptations to aerobic exercise training. Sports Med 1989; 8: 302-321.

Nieman DC. Exercise and upper respiratory tract infection. Sports Med Training Rehabil 1993; 4: 1-14.

Niess AM, Dickhuth HH, Northoff H, et al. Free radicals and oxidative stress in exercise—immunological aspects. Exer Immunol Rev 1999; 5: 22-56.

Noakes TD. Physiological models to understand exercise fatigue and the adaptations that predict or enhance athletic performance. Scand J Med Sci Sports 2000; 10: 123-145.

Noble BJ, Robertson RJ. Perceived Exertion. Champaign, IL: Human Kinetics, 2000.

O'Connor PJ, Morgan WP, Raglin JS, et al. Mood state and salivary cortisol levels following overtraining in female swimmers. Psychoneuroendocrinology 1989; 14: 303-310.

Ostrowski K, Rohde T, Asp S, et al. Pro- and anti-inflammatory cytokine balance in strenuous exercise in humans. J Physiol 1999; 515: 287-291.

Papoti M, Martins LEB, Cunha SA, et al. Effects of taper on swimming force and swimmer performance after an experimental ten-week training program. J Strength Cond Res 2007; 21: 538-542.

Peake JM. Exercise-induced alterations in neutrophil degranulation and respiratory burst activity: possible mechanisms of action. Exer Immunol Rev 2002; 8: 49-100.

Pedlar C, Whyte G, Emegbo S, Stanley N, Hindmarsh I, Godfrey R. Acute sleep responses in a normobaric hypoxic tent. Med Sci Sports Exerc 2005; 37: 1075-1079.

Pizza FX, Flynn MG, Boone JB, et al. Serum haptoglobin and ferritin during a competitive running and swimming season. Int J Sports Med 1997; 18: 233-237.

Pizza FX, Mitchell JB, Davis BH, et al. Exercise-induced muscle damage: effect on circulating leukocyte and lymphocyte subsets. Med Sci Sports Exer 1995; 27: 363-370.

Prins JH, Lally DA, Maes KE, et al. Changes in peak force and work in competitive swimmers during training and taper as tested on a biokinetic swimming bench. In: Aquatic Sports

Medicine. JM Cameron, Ed. London: Farrand Press, 1991: 80-88.

Pyne DB, Gleeson M, McDonald WA, et al. Training strategies to maintain immunocompetence in athletes. Int J Sports Med 2000; 21: S51-S60.

Pyne D, Mujika I, Reilly T. Peaking for optimal performance: research limitations and future directions. J Sports Sci 2009; 27: 195-202.

Raglin JS, Koceja DM, Stager JM. Mood, neuromuscular function, and performance during training in female swimmers. Med Sci Sports Exerc 1996; 28: 372-377.

Raglin JS, Morgan WP, O'Connor PJ. Changes in mood states during training in female and male college swimmers. Int J Sports Med 1991; 12: 585-589.

Reilly T, Atkinson G, Budgett R. Effects of low-dose temazepam on physiological variables and performance following a westerly flight across five time-zones. Int J Sports Med 2001; 22: 166-174.

Reilly T, Atkinson G, Edwards B, et al. Coping with jet-lag: a position statement of the European College of Sport Science. Eur J Sport Sci 2007a; 7: 1-7.

Reilly T, Edwards B. Altered sleep-wake cycles and performance in athletes. Physiol Behav 2007; 90: 274-284.

Reilly T, Maskell P. Effects of altering the sleep-wake cycle in human circadian rhythms and motor performance. In: Proceedings of the First IOC World Congress on Sport Science. Colorado Springs, CO: U.S. Olympic Committee, 1989: 106.

Reilly T, Waterhouse J, Burke LM, Alonso JM. Nutrition for travel. J Sport Sci 2007b 25: S125-S134.

Reilly T, Waterhouse J, Edwards B. Jet lag and air travel: implications for performance. Clin Sport Med 2005; 24: 367-380.

Rietjens GJWM, Keizer HA, Kuipers H, et al. A reduction in training volume and intensity for 21 days does not impair performance in cyclists. Br J Sports Med 2001; 35: 431-434.

Rudzki SJ, Hazard H, Collinson D. Gastrointestinal blood loss in triathletes: its etiology and relationship to sports anaemia. Aust J Sci Med Sport 1995; 27: 3-8.

Rushall BS, Busch JD. Hematological responses to training in elite swimmers. Can J Appl Sports Sci 1980; 5: 164-169.

Rusko H, Tikkanen HO, Peltonen JE. Altitude and endurance training. J Sport Sci 2004; 22: 928-945.

Sahlin K, Broberg S. Adenine nucleotide depletion in human muscle during exercise: causality and significance of AMP deamination. Int J Sports Med 1990; 11: S62-S67.

Saltin B, Gollnick PD. Skeletal muscle adaptability: significance for metabolism and performance. In: Handbook of Physiology: Skeletal Muscle. Bethesda, MD: American Physiological Society, 1983: 555-631.

Seiler D, Nahel D, Franz H, et al. Effects of long-distance running on iron metabolism and hematological parameters. Int J Sports Med 1989; 10: 357-362.

Selby GB, Eichner ER. Endurance swimming, intravascular hemolysis, anemia, and iron depletion. Am J Med 1986; 81: 791-794.

Shepley B, MacDougall JD, Cipriano N, et al. Physiological effects of tapering in highly trained athletes. J Appl Physiol 1992; 72: 706-711.

Smith HK. Ergometer sprint performance and recovery with variations in training load in elite rowers. Int J Sports Med 2000; 21: 573-578.

Smith LL. Overtraining, excessive exercise, and altered immunity: is this a T helper-1 versus T helper-2 lymphocyte response. Sports Med 2003; 33: 347-364.

Smith JA, Pyne DB. Exercise, training and neutrophil function. Exer Immunol Rev 1997; 3: 96-117.

Smith LL, Anwar A, Fragen M, et al. Cytokines and cell adhesion molecules associated with high-intensity eccentric exercise. Eur J Appl Physiol 2000; 82: 61-67.

Snyder AC, Jeukendrup AE, Hesselink MKC, et al. A physiological/psychological indicator of over-reaching during intensive training. Int J Sports Med 1993; 14: 29-32.

Stein M, Keller SE, Schleifer SJ. Stress and immunomodulation: the role of depression and neuroendocrine function. J Immunol 1985; 135: 827-833.

Steinacker JM, Lormes W, Kellmann M, et al. Training of junior rowers before world championships. Effects on performance, mood state and selected hormonal and metabolic responses. J Sports Med Phys Fitness 2000; 40: 327-335.

Stewart AM, Hopkins WG. Seasonal training and performance in competitive swimmers. Med Sci Sports Exerc 2000; 32: 997-1001.

Stone MH, Josey J, Hunter G, et al. Different taper lengths: effects on weightlifting performance. In: Proceedings of the Overtraining and Overreaching in Sport International Conference 1996; University of Memphis, Tennessee, p. 59.

Suzuki K, Nakaji S, Yamada M, et al. Systemic inflammatory responses to exhaustive exercise: cytokine kinetics. Exer Immunol Rev 2002; 8: 7-48.

Tabata I, Irisawa K, Kouzaki M, Nishimura K, Ogita F, Miyachi M. Metabolic profile of high intensity intermittent exercises. Med Sci Sports Exerc 1997; 29: 390-395.

Taha T, Thomas SG. Systems modelling of the relationship between training and performance. Sports Med 2003; 33: 1061-1073.

Tanaka H, Costill DL, Thomas R, et al. Dry-land resistance training for competitive swimming. Med Sci Sports Exerc 1993; 25: 952-959.

Taylor SR, Rogers GG, Driver HS. Effects of training volume on sleep, psychological, and selected physiological profiles of elite female swimmers. Med Sci Sports Exerc 1997; 29: 688-693.

Telford RD, Hahn AG, Catchpole EA, et al. Postcompetition blood lactate concentration in highly ranked Australian swimmers. In: Swimming Science V. Champaign, IL: Human Kinetics, 1988: 277-283.

Tharp GD, Barnes MW. Reduction of saliva immunoglobulin levels by swim training. Eur J Appl Physiol 1990; 60: 61-64.

Tharp G, Preuss T. Mitogenic response of T-lymphocytes to exercise training and stress. J Appl Physiol 1991; 70: 2535-2538.

Thomas L, Busso T. A theoretical study of taper characteristics to optimise performance. Med Sci Sports Exerc 2005; 37: 1615-1621.

Thomas L, Mujika I, Busso T. A model study of optimal training reduction during pre-event taper in elite swimmers. J Sport Sci 2008; 26: 643-652.

Thomas L, Mujika I, Busso T. Computer simulations assessing the potential performance benefit of a final increase in training during pre-event taper. J Strength Cond Res; in press.

Trappe S, Costill D, Thomas R. Effect of swim taper on whole muscle and single fiber contractile properties. Med Sci Sports Exerc 2001; 33: 48-56.

Trinity JD, Pahnke MD, Resse EC, Coyle EF. Maximal mechanical power during a taper in elite swimmers. Med Sci Sports Exerc 2006; 38: 1643-1649.

Tullson PC, Terjung RL. Adenine nucleotide metabolism in contracting skeletal muscle. Exerc Sports Sci Rev 1991; 19: 507-537.

Van Handel PJ, Katz A, Troup JP, et al. Oxygen consumption and blood lactic acid response to training and taper. In: Swimming Science V. Champaign, IL: Human Kinetics, 1988: 269-275.

Verde TJ, Thomas S, Shek P, et al. Responses of lymphocyte subsets, mitogen-stimulated cell proliferation, and immunoglobulin synthesis to vigorous exercise in well-trained athletes. Clin J Sports Med 1992; 2: 87-92.

Viru A. Plasma hormones and physical exercise. Int J Sports Med 1992; 13: 201-209.

Viru A, Viru M. Biochemical Monitoring of Sport Training. Champaign, IL: Human Kinetics, 2001.

Vollaard NB, Cooper CE, Shearman JP. Exercise-induced oxidative stress in overload training and tapering. Med Sci Sports Exerc 2006; 38: 1335-1341.

Wade CH, Claybaugh JR. Plasma renin activity, vasopressin concentration, and urinary excretory responses to exercise in men. J Appl Physiol 1980; 49: 930-936.

Walker JL, Heigenhauser GJF, Hultman E, et al. Dietary carbohydrate, muscle glycogen content, and endurance performance in well-trained women. J Appl Physiol 2000; 88: 2151-2158.

Warren BJ, Stone MH, Kearney JT, et al. Performance measures, blood lactate and plasma ammonia as indicators of overwork in elite junior weightlifters. Int J Sports Med 1992; 13: 372-376.

Waterhouse J, Atkinson G, Edwards B, Reilly T. The role of a short post-lunch nap in improving cognitive, motor and sprint performance in participants with partial sleep deprivation. Journal of Sports Sciences 2007a; 25: 1557-1566.

Waterhouse J, Nevill A, Edwards B, Godfrey R, Reilly T. The relationship between assessments of jet-lag and some of its symptoms. Chronobiol Int 2003; 20, 1061-1073.

Waterhouse J, Reilly T, Atkinson G, Edwards B. Jet lag: trends and coping strategies. *Lancet* 2007b; 369: 1117-1129.

Watt B, Grove R. Perceived exertion: antecedents and applications. Sports Med 1993; 15: 225-241.

Weight LM. "Sports anaemia": does it exist? Sports Med 1993; 16: 1-4.

Wenger HA, Bell GJ. The interactions of intensity, frequency and duration of exercise training in altering cardiorespiratory fitness. Sports Med 1986; 3: 346-356.

Wilson M, Kreider R, Ratzlaff R, et al. Effects of a 3-week taper period following 22-weeks of intercollegiate swim training on fasting immune status. In: Proceedings of the Overtraining and Overreaching in Sport International Conference 1996; University of Memphis, Tennessee, p. 73.

Yamamoto Y, Mutoh Y, Miyashita M. Hematological and biochemical indices during the tapering period of competitive swimmers. In: Swimming Science V. Champaign, IL: Human Kinetics, 1988: 269-275.

Zarkadas PC, Carter JB, Banister EW. Modelling the effect of taper on performance, maximal oxygen uptake, and the anaerobic threshold in endurance triathletes. Adv Exp Med Biol 1995; 393: 179-186.

Note: The italicized *f* and *t* following page numbers refer to figures and tables, respectively.

Photo courtesy of Iñigo Mujika

Iñigo Mujika, PhD, is a sports physiologist at USP Araba Sport Clinic in Vitoria-Gasteiz, Basque Country, and an Associate Professor at the Department of Physiology, Faculty of Medicine and Odontology, at the University of the Basque Country. Previously, Mujika was also the head of the department of research and development at the Spanish professional football team Athletic Club Bilbao, Basque Country. As a researcher, sport science practitioner, and coach, Mujika is widely considered one of the most respected experts on tapering and peaking for optimal performance.

Since 1992, Mujika has been devoted to the research of tapering and peaking for sport performance. He has published over 20 peer-reviewed scientific articles, 6 book chapters, and 10 other publications on tapering-related issues. He has also presented nearly 70 lectures on tapering at conferences and seminars worldwide.

As a sport physiologist, Mujika works closely with elite athletes and coaches in a variety of individual and team sports. From 2003 to 2004, Mujika was senior physiologist at the Australian Institute of Sport. In 2005, he worked as physiologist and trainer of the professional road bicycle racing team Euskaltel Euskadi. He is also a coach of world-class triathletes, having coached Olympians Ainhoa Murua to Athens 2004 and Beijing 2008, and Eneko Llanos to Athens 2004.

Mujika is a member of the American College of Sports Medicine and serves as associate editor of the *International Journal of Sports Physiology and Performance*. He holds a doctoral degree in biology of muscular exercise (1995) from the University of Saint-Etienne in France and a second doctorate in physical activity and sport sciences (1999) along with an Extraordinary Doctorate Award from the University of the Basque Country. In 2002 and 2007 he received the National Award for Sport Medicine Research from the University of Oviedo in Spain. He has also received two awards for his work with triathletes: Best Coach of Female Athlete (2006) from the Spanish Triathlon Federation and the High Performance Basque Sport Award (2007) from the Basque Sport Foundation.

Fluent in four languages (Basque, English, French, and Spanish), Mujika has lived in California, France, South Africa, and Australia. He currently resides in the Basque Country. Mujika enjoys surfing, cycling, swimming, strength training, and hiking, as well as cinema and traveling.